Transient Workspaces

Transient Workspaces

Technologies of Everyday Innovation in Zimbabwe

Clapperton Chakanetsa Mavhunga

The MIT Press
Cambridge, Massachusetts
London, England

This book was set in ITC Stone Serif Std by Toppan Best-set Premedia Limited.

Library of Congress Cataloging-in-Publication Data
Mavhunga, Clapperton Chakanetsa, 1972–
Transient workspaces : technologies of everyday innovation in Zimbabwe / Clapperton Chakanetsa Mavhunga.
    pages   cm.—(Mobility studies)
Includes bibliographical references and index.
ISBN 978-0-262-02724-3 (hardcover : alk. paper) | 978-0-262-53758-2 (pbk.) 1. Material culture—Africa. 2. Subsistence hunting—Africa. 3. Economic anthropology—Africa. I. Title.
GN645.M3526 2014
306.4'6096—dc23
2013036289

For the reader of footprints, tracker of people and animals, whose technologies and innovations are now criminalized as poaching.

Contents

Acknowledgments

In the chiShona language of Zimbabwe, to not say *"Ndatenda"* (Thank you) is not just an indictment of one's humanity. *Kusatenda uroyi* (To not thank is witchcraft). I am not a witch, so I will say *Ndatenda* to many people who have shaped what I do in this book.

First, I wish to thank two professors on the knowledge of my ancestors: my father Peter Masango Mavunga and my mother Violet Mavunga, née Chiseka. This book is following the footprints of your wisdom. May I be a good steward of it and pass it on to my children. To my late brother Denny; our strict yet loving mentor in the valleys, rivers, forests, and fields in which we learned and mastered herding, fishing, trapping birds with rubber (urimbo) you taught us to make from the chitatarimbo tree. I will always love and miss you.

Ndatenda vaKusemamuriwo, my Form 1 chiShona teacher at Chizengeni Secondary School; vaMuparamoto, my O-level chiShona teacher at Rakodzi High School; and Amai Chiunda, my A-level chiShona teacher at Marondera High School. And to my English teacher at Rakodzi, Bernard Matesanwa, thank you for inspiration.

I also thank Arwen Mohun at University of Delaware, who referred me to Gabrielle Hecht when I wanted to start studying science, technology, and society (STS) in the context of Africa—which I eventually did, at the University of Michigan. To Gabrielle Hecht, an exceptional doctoral adviser, a friend indeed—what a wonderful human being, advisor, and teacher!—thank you, *ndatenda*. My thanks also to Nancy Rose Hunt, who first welcomed me to Michigan, and shaped my *longue durée* approach to things and the need to appreciate the work that Africanists have done more closely and not get carried away with STS. Zingerman's was the place where our Africa reading group would meet for hours! To David William Cohen, who encouraged us to think about the politics of knowledge production in Africa. To Mamadou Diouf, who introduced me to African philosophy, and to whom I owe the commitment to questions of knowledge that matter to who we are as Africans. To Arun Agrawal and Rebecca Hardin, two mentors on matters environmental. To Kathleen Canning and Michelle Mitchell—what exceptional professors of "that!" introductory history class of fall 2003! And to dear Wolverine friends whom I counted on when the chips were down—Sarah Hillewaert, Menna Demessie, Christian Williams, Heloise Finch, Doris Essah, Sudipa Topdar, Stephen Sparks, and Nafisa Essop. May your hearts be blue forever!

To the anonymous reviewers of this manuscript and the editorial, publicity, and design teams at the MIT Press—Marguerite Avery, Kate Persons, Kathleen Caruso, Julia Collins, Sarah Courtney, Susan Clark, and Sean Reilly—thank you.

It is five years now since I arrived at MIT. I have been fortunate to receive advice from Roe Smith, Rosalind Williams, David Mindell, and Dave Kaiser. My thanks to Natasha Schüll, Hanna Shell, and Vincent Lepinay, for the wonderful companionship of junior faculty. To Katrien Pype, whose energy for common intellectual pursuits inspired me so much. To Mhoze Chikowero, for the long Skype conversations from Santa Barbara.

I would like to thank the people of Chibwedziva, Malipati, Mahenye, and Makuleke for opening their doors to educate me on their knowledge. *Nikensile ngofu* (thank you so much) to my research team of Elmon Magezi Chauke, Attorney Hlongwane, the late Benes Maluleke, and Solomon Bvekenya, you are truly indigenous knowledge archives on two feet. A special thanks to librarians at the National Archives of Zimbabwe, the MIT Library, and Cullen Library at Wits for helping me find archival materials. I am especially indebted to Senior Ranger Jafta Chikwanha and Tovadini Kemusi for taking me around Gonarezhou and allowing me to use the archival library at Mabalauta, and to interview game rangers and veterans of the Zimbabwe war of liberation. Since then, the research that Solomon Bvekenya has led locally has evolved and is on the verge of great things.

This book also profited from many generous invitations to conferences and workshops, to present public lectures, and to be a fellow in residence. I wish to thank Deborah Posel, Sarah Nuttall, Robert Muponde, Tom Odhiambo, and Najibha and Adila Deshmukh for the memorable and productive time I spent in residence at the Wits Institute for Social and Economic Research (WISER) in 2005. I spent six months in paradise at the Rachel Carson Center in 2011 completing one of two manuscripts. To its directors Christof Mauch and Helmuth Trischler, thank you for the generous financial support, smooth logistics, and incomparable intellectual space. To my fellow fellows Dorothee Schreiber, Riin Magnus, Maohong Bao, Lawrence

Culver, Fiona Cameron, Marianna Dudley, Ingo Heidbrink, Timothy LeCain, Istvan Praet, Gary Martin, and Jagdish Dawar—I profited from your comments and wonderful company.

I wish to acknowledge the abiding inspiration of Aaron Hodza, the late Gordon Chavhunduka, Ngugi wa Thiong'o, V. Y. Mudimbe, Paulin Hountondji, Mamadou Diouf, Mahmood Mamdani, Manthia Diawara, Achille Mbembe, Chinua Achebe, and Kwasi Wiredu, whose ideas continue to remind some of us of the work yet to be done to write an Africa-centered narrative.

I wish to thank the following schools and programs for inviting me to present various incarnations of my chapters: the Walter Rodney Seminar at Boston University; STS Circle at Harvard; Virginia Tech STS; African Studies and Sawyer seminars at Indiana University Bloomington, in particular Marissa Moorman and Eden Medina; Cornell STS; the Joint DPP, IKD & INNOGEN Seminar at Open University (UK); the Johannesburg Workshop on Theory and Criticism; Africana Studies Program at Bard College; Ruhr-Universität Colloquium on Early Modern Environmental History; the Rachel Carson Center's Lunchtime Colloquium; and the Program in Science, Technology, Environment and Health at Rutgers.

Over the past decade this research has received generous support from various sources at the University of Michigan and at MIT, for which I am exceedingly grateful. At Michigan I received the Institute for Research in Women and Gender (IRWIG) Travel Grant (2004), the International Institute's Pre-dissertation Research Award (2004), the African Initiative Travel Grant (2004 and 2006), the South Africa Initiative Travel Grant (2004 and 2006), the Rackham Humanities Research Candidacy Fellowship (2006–2007), the Rackham International Research Award (2006–2007), the Melvin and Janey Lack Fellowship (2006–2007), the Global Ethnic Literatures Seminar (GELS) Fellowship (2007–2008), and the Rackham Dissertation Fellowship (2007–2008). At MIT I am grateful as well for support from the SHASS Research Fund (2011), the Dean's Faculty Development Fund (2013), and the Program in Science, Technology, and Society.

Ndatenda with all my heart to Mildred, and to our daughter Nyasha and sons Chitsidzo-Promise and Dadiso Pride, for tolerating the time I was away from you while writing this book, for offering love often unreturned, and for having confidence in what I do even when I have none left in me— I thank you.

C.C.M., Framingham, Massachusetts, USA, May 30, 2013

Prologue

On Wednesday, September 25, 2013, Zimbabwe's newly appointed Minister of Environment Saviour Kasukuwere announced to the nation that "poachers" had massacred more than 90 elephants in one stroke in the country's largest game reserve, Hwange National Park. The massacre occurred at a saltpan inside the game reserve, adjacent to villages in the Pelandaba area of rural Tsholotsho. Rangers tracked the spoor from the stench-filled pool to the homestead of one of the suspects, Sipho Mafu, who admitted to having committed the crime with two accomplices (Mudzungairi 2013). In a subsequent sweep of the villages, police and rangers of the Zimbabwe Parks and Wildlife Management Authority (ZPWMA) recovered nineteen tusks, barrels full of cyanide, and wire snares ("Zimbabwe Poachers" 2013).

What shocked the government was not only the numbers killed at once, or that these were ordinary people in villages along the national park's boundaries doing it, but also the "sophisticated" technology they were using. For more than a century, the government had come to associate "poaching" with wire snares and firearms—"the traditional way of poaching" (Cox 2013). There had also been sporadic cases in which villagers laced oranges and watermelons with agricultural pesticides to kill rhinos, and sometimes elephants, but these methods targeted individual animals, not entire herds. The snare killed only one animal entering through the noose, the arrow and the bullet one animal per shot. Cyanide destroys indiscriminately and en masse. All animals drinking the water, living in pools, licking salt at the saltpans, or eating the poisoned carcasses die—entire herds of elephant, prides of lions, packs of wild dogs, schools of hippos, troops of baboons, flocks of birds, and even, one day, the villagers themselves. As the summer rains arrive, the flooded rivers carry the cyanide downstream toward the Zambezi River, and off to sea to poison sea life too—and seafood lovers (Magadza 2013; Mukarati 2013).

Cyanide is a family of hydrogen cyanide salts that contain the ion cyanide (CN, as distinct from the chemical element copernicium, Cn). The family includes the highly toxic compounds potassium cyanide and sodium cyanide. The chemical kills by shutting down cellular respiration, inducing the body into a coma, seizures, cardiac arrest, and death within minutes. Cyanide poisoning occurs when a living organism comes into contact with the compound and the cyanide ions dissolve in water (Mukarati 2013). It is photodegradable, meaning eventually it will lose its potency with time, but the carnage to wild animals will already be immense.

Of course, villagers engaged in poaching do not have to know that minutiae to be able to use the chemical. All they need to know, and care about, is that cyanide has lethal properties that enable them to kill a giant animal and harvest its tusks and, through them, earn a living. Villagers acquire the cyanide through their contacts, who work for chemical companies in the city. When they get the cyanide drums from the suppliers, they hide them underneath the grain huts, in the bush, and in caves or bury them in the ground. On the day of the killing, they take only enough cyanide for their designated target and stow away the rest for future use. The poachers basically use two methods of application. One is to poison the saltpans where elephants go to lick salt and to drink water; the other is to place buckets of poisoned water along the elephant trails leading to the water hole since the animals use the same route each time. They then monitor the animals' movements, knowing that the poison takes effect within minutes ("Cyanide Poachers" 2013). The poachers ensure that they post sentinels in the hills or big trees to watch the landscape below for patrols. They then move in with speed, cut out the tusks from the animals' heads, and lie low until the sun sets, before exiting the park under cover of darkness.

When they get to the main road leading to the nearest town called Tsholotsho Growth Point, the poachers use donkeys, usually stolen in other villages, which they then set free to roam, far from their owners, once they have done their job ("Cops Attacked" 2013). At Tsholotsho the poachers use private vehicles or *amalayitsha* (public commuter omnibuses) to transport the tusks to Zimbabwe's big cities, Harare or Bulawayo. They surrender the ivory to the dealers—who may well be top politicians, top cops, Zimbabwean or foreign businesspeople, and even clergymen—in return for US$200 to US$500 per pair of tusks ("We Poisoned Jumbos" 2013).

These "sharks" then sell each tusk in Asia and the Middle East for around US$17,000. The ivory is smuggled out of the country either directly via the Harare International Airport to Dubai, Thailand, and China, or by road into neighboring Zambia and South Africa. The syndicates have implicated officials within the ZPWMA, police, and immigration, who ensure that the tusks pass through police roadblocks (traffic stops) and customs and immigration control points without problems (Muponde 2013; "We Poisoned Jumbos" 2013). In Thailand and China, the ivory is carved into jewelry, spear tips, buckles, piano keys, sword handles, clothing buttons, billiards or snooker balls, combs, hairpins, and chopsticks. Alongside rhino horn, ivory is a key ingredient in a range of Chinese traditional medicines, including aphrodisiacs (Mukarati 2013; "447 kg of Ivory" 2013). Almost 80 percent of the world's population relies on traditional medicines, with China's by far the most sought after in Asia and increasingly in Africa. The ivory poached in the national park may well be returning to Zimbabwe as an ingredient in this powdered medicine (WHO Fact Sheet No. 134 2003; "Kangai" 2012).

Introduction

The problem with the national park or game reserve in Africa is that it is a colonial relic struggling to adjust to a postcolonial reality. Established in 1928 as a game reserve and proclaimed a national park (state-owned game reserve) in 1961, Hwange is an example of how people were violently removed from their homelands of choice and dumped in arid, infertile, and pest-infested peripheries to make way for nature conservation. At the pain of arrest or being shot, they were banned from entering the game reserve except as cooks, game scouts, and general hands, even as colonial settlers and tourists from Europe and North America came in to enjoy the game drives and the hunting safaris. To villagers displaced to its periphery, the game reserve was a most sadistic colonial act.

At independence, sub-Saharan African governments inherited and perpetuated the game reserve in its colonial form, albeit now with black rangers and scouts. The sprawling size, growing animal populations, and an ever-shrinking budget for its upkeep has placed severe constraints on the continued maintenance of the game reserve as a "fortress" against villagers. The Department of National Parks and Wild Life Management (DNPWLM) and its successor, the Zimbabwe Parks and Wildlife Management Authority (ZPWMA), has always struggled with staffing, funding, and equipment since its beginnings. But the problem has worsened as the economy weakened in the last decade. Today one game ranger patrols an average radius of 125 miles compared to the standard thirteen miles. Zimbabwe needs no less than US$40 million per annum to effectively manage its 120,000-strong elephant herd, to say nothing of other animals, against the background of other urgent economic priorities ("Hwange Clean-up" 2013). It is understandable, therefore, that some conservationists are tempted to suggest the purchase and deployment of drones to equip the game reserve with a capacity for satellite-guided surveillance (Magadza 2013).

Others are not so sure that *yet more technology* is enough so long as the coercive model of conservation remains in place. Even while "declaring war on the poachers" and promising that the government will respond "with all our might," Environment Minister Saviour Kasukuwere has warned that without addressing the grievances of villagers, poaching will be "difficult, if not impossible, to eradicate" (Chara 2013). The minister added that when people see no benefits from what is supposed to be their resource, "they will work with foreigners to destroy their birthright. . . . Tsholotsho is an arid area that has been hit by a series of poor harvests; it only makes sense that people here should benefit from their wildlife" ("We Poisoned Jumbos" 2013).

The notion of "benefit" is a rather nebulous one. Since 1987 the villagers had been working with Communal Areas Management Programme for Indigenous Resources (CAMPFIRE), a donor-funded program that seeks to involve communities surrounding nature reserves in the management (but not ownership) of forest resources. In return for giving up any access to the animal, plant, or cultural resources inside conservancies, villagers would be given meat and a percentage of earnings from trophy hunting, which only hunters from overseas can afford. Interestingly, CAMPFIRE has depended almost entirely on donor funding from the same Western countries that these hunters come from.

A note of caution on the word "poach," its derivative "poaching," and the "game" associated with it. These are words that are deeply problematic and colonial. "Poach" as opposed to "hunt" clearly has its English roots in the word "pocchen" or "bagged." Initially, the person who bagged was personified in the romantic figure of Robin Hood, of whom it is "not to be inquired, whence comes the venison" (Schama 1995, 137). Then, in the nineteenth century, the meaning and practice of poaching was increasingly criminalized. Laws such as the Night Poaching Act (1828) and the Game Act (1831) were passed, at just the time when British citizens began arriving in southern Africa, some to settle permanently. They brought in the idea of "game" as an animal that exists for human amusement or purposes, that is, to be bagged. The African increasingly became the poacher, the one whose accessing of game could only be criminal. The meanings of animals were now reduced to the secular, the culinary, the aesthetic, and the financial. Forest animals (*mhuka*) were de-spiritualized and dehumanized into mere fauna or game, the sacred groves become just flora, African access to both regulated and criminalized. Therefore, the decision not to scare quote them was editorial; whenever the reader encounters them, they should treat them as "the colonizer's language," not mine or others of Africans of whom I write. The use of "forest animals" and "hunting" as opposed to game and poaching is meant to restore the spiritualized and humanized values of the forest and its animals according to African values. It is a decolonializing move.

In a sense, "poaching" is a critique of a top-down view of "benefit" defined and introduced from outside without involving ordinary people living in the proximity of the national park at the stage of conception. Ordinary people in the rural countryside often complain that the political and financial powers of politicians and international nongovernmental organizations (NGOs) reduce them to people who can only "hear but cannot speak" (Mavhunga and Dressler 2007). The effect of these unidirectional power relations is to alienate biodiversity conservation from the very people who live with and might act as its first line of defense from unscrupulous outsiders and corrupt politicians. In development projects, ordinary people have no space for contributing ideas that might improve programs that are implemented in their name. People particularly resent being consulted at implementation, but never in the conception of projects.

Development continues to be *for them*, but not *with them* (Tapela, Maluleke, and Mavhunga 2007). The much-touted "partnership" between rich and poor countries, politicians and "their" people, (non)governmental organizations and villagers, overseas and local universities, and the university and communities, becomes merely a partnership between a rider and a horse, the one enjoying the ride and directing the itinerary, the other shouldering the burden and doing as ordered (Mavhunga 2007b). This is why development, conservation, technology, and innovation projects fail, not only in Africa or the global south, but universally. The view of partnership as a relationship between "us" as riders and ordinary people as horses also forecloses a view to ordinary people as creative beings in their own right.

Technologies of Everyday Innovation

The cyanide poacher is neither a freak nor a sudden development, but is heir to generations of hunting traditions that villagers have from time to time adapted to help themselves to newly available means of killing animals. Nor is it just hunters; African history is replete with examples of people adjusting their traditions to craft self-help solutions to everyday challenges and to selectively tap into resources from outside.

This book seeks to historicize poaching as a historical example of the means and ways with which ordinary people engage in creative activities directed toward solving their problems and generating values relevant to their needs and aspirations. Throughout this book, "innovation" means the act of introducing something new, be it a method or a thing, either from scratch or from outside. Even in instances where that which ordinary people innovate upon and with is coming from outside, this book draws attention to the capacities people already have that enable them to import and deploy it. The people are seen in conversation with the "interpretively flexible" material (Pinch and Bijker 1984) of the incoming thing and assigning it new meanings and purposes.

The intention of *Transient Workspaces* is therefore to use the hunt in Zimbabwe as a narrow path to a larger dialogue on the place of ordinary people in technology. Given a choice between seeing them merely as recipients of technology and innovation and treating them as designers and innovators in their own right, the book takes the latter position, starting from African vernaculars to establish dialogue with the designer-user interfaces explored in the works of Pinch and Bijker (1984), Woolgar (1991), Oudshoorn and Pinch (2003), and Edgerton (2007). Thus instead of tracing the journey of cyanide and firearms from their "designers" overseas to their "users" in Zimbabwe, I see *the villagers as the designers* using cyanide as a resource to turn a large mammal into ivory for sale to markets and users in Asia and the rest of the world.

Indeed, one headline in the aftermath of the massacre suggested that the use of the chemical "reveal[s] alarming innovation in poaching tactics" (Cox 2013). Shocking as it may seem, the use of cyanide as a weapon of mass animal destruction is not a new phenomenon in grassroots innovations that Africans bring to bear upon incoming things. Those of us who grew up in rural Africa see the home, the village, the mountains, the valleys, and the rivers as educational and technological spaces where these innovations occurred on a daily basis. Such spaces are indeed the universities, the laboratories, or the factories of psychomotor activities within which many of Africa's leaders in politics, business, academia, sport, music, and many other endeavors are raised. The valley where children herd cattle, the pools where they fish, the forests in which they hunt and pick fruit, the dusty streets where they play with their self-made plastic football—all these are sites in which the African child is taught critical life skills through showing and doing, but not the exam or the pen. Out of them arises a spirit of experiment, adventure, risk-taking, and ambition, inspired by a desire to escape grinding realities of being born poor through sheer hard work and seeking answers in novelty even while one's feet are firmly planted in the elastic cultural traditions of one's ancestors.

It is in times of economic and political crisis that these technologies of the ordinary people manifest themselves ever more boldly. This is how they survived the economic crisis in Zimbabwe that peaked in 2008, a time when supermarket shelves were empty, fuel stations were dry, and hyperinflation reached 79,600,000 percent per month, and government was totally powerless to do anything. It is not a coincidence that the most common feature of Zimbabwean life during that period was mobility: of people moving abroad to work in all kinds of jobs, people walking tens of miles on foot to work or between cities and the countryside because fuel imports had dried up, people traveling within the southern Africa region buying and selling "stuff," people waiting in queues to buy scarce commodities—and indeed the sheer effort of keeping the body itself moving in defiance of the hunger, diseases, and stress.

Many of the activities that ordinary people engaged in to sustain themselves were illegal under the law: for example, mobile street vending (as opposed to designated market buildings like Mbare Musika, Chikwanha, or Dombotombo), foreign currency trading (always on foot touting and bargaining), pirating (unlicensed shuttle services using one's personal car to make money)—the list of "informal" sector entrepreneurship was endless. It is from such unlikely places as the scene of poaching that this book is written and where questions about Africa in technology are addressed.

Mobility and Technologies of Everyday Innovation

How then does one write, let alone go about finding, innovation in such a grotesque and putrid site as the cyanide massacre in Hwange National Park? In a practice—poaching—that one finds morally repugnant and that offends one's sensibilities? In a practice that the colonial state criminalized through the law and the postcolonial state retains? How does one find innovation in Africa by Africans for themselves, in a global semantic environment where technology is generally perceived to be "transferred" from the north, and innovation is something that outsiders bring to Africa? In a remote rural village where no roads, electricity, or cell phone connectivity exists, and the movements, sights, and sounds of francolins, hyenas, elephant, and lions animate the diurnal and nocturnal atmosphere?

Telling the story of everyday innovation from places like these, located on the borderlands of Zimbabwe, is all the more interesting because they exemplify in the Western imaginary the antithesis of technology and innovation. In such a Western-centric view, everything that is good on and about Africa came from outside: art from the Portuguese, architecture from

Phoenicians and Arabs, statecraft from Europe, crop and animal domestication from southwest Asia, astronomy from the Romans—the list goes on.
One can't help but remark: Africa's indebtedness to outsiders is unredeem-
able (Mudimbe 1988, 13–15). Indeed, history has dealt Africa a heavy hand,
from the time of the trans-Atlantic slave trade, to colonialism, to postcolonial economic structural adjustment programs that the International
Monetary Fund (IMF) and World Bank imposed.

How could one expect to find technology and innovation (by Africans)
amid all this "chaos"—of war, "disease," famine, poverty, dictatorships, and
corruption? On a continent governed by forces beyond its control? A sick
continent whose signatures of technology are the begging bowl, the Red
Cross symbol, and SUVs belonging to nongovernmental organizations and
corrupt local elites? Among images of *poverty* buried in "hordes of vital
statistics" (Fanon 1952/1967, 33; Kanbur and Squire 2001; Green and
Hulme 2005, 867) that simplify the problems of Africa for NGOs trying to
design solutions? Amid a grammar of victimhood that excludes any other
states of and potential for being and becoming outside a psychology of
helplessness, etched in memory and imaginable but not imaginative
future(lessness)? This reminder of the "dark" in *Heart of Darkness* (Conrad
1902) posits Africa as "a mere bad copy, a negative image of someone else's
positivity" (Njamnjoh 2004, 317). Impelled by slavery, colonialism, and
apartheid into this lifeless "objecthood," the negativist's Africa becomes
not only terra incognita to the Other, but loses even the sense of self-rec-
ognition (Mbembe 2002, 243). Such are the magical powers of the Other's
inscription devices that Africa counts only as an invention.

Indeed, in the narratives of globalization, Africa appears as a technological and cultural artifact, one more sign of the irresistible powers of capitalist
invention. The World Wide Web completes the work of networking this
vast empire of capital that newspapers, radio, and television had sustained
since the Cold War. So insignificant is Africa in the globalization narrative
that, as James Ferguson (2006) observed, prominent scholarship could
easily talk about "the world" while saying almost nothing at all about
Africa (Castels 1996; Sassen 1999; Stiglitz 2003; Hardt and Negri 2001;
Giddens 2002; Paolini 1997; Held et al. 1999).

Why, after all, does one need to worry about Africa? It has contributed
nothing to that which marks out the contours of our "modern" existence—
technology, especially the sort that enables us to move information, goods,
and people, and to connect fast and in real time. In any case, if Africa's
contribution to these things that mark world time really mattered or

existed, wouldn't its historians, anthropologists, political scientists, engineers, and scientists have written about it already? They have not, and we can safely assume they have nothing to write about Africa that resides in the domain of technology and innovation except how the technological innovations of others have victimized them. That, or how Africans are "reacting" to incoming technologies, or how such technologies are impacting them, is all they write about. If the direction from which one can approach Africa in the narrative of "modernity" is only as a victim or simply reacting, then Africa reflects a lack of initiative.

The complaint has been made that these macrocultural approaches to globalization and modernity have no "sensitivity to cultural production at a local or regional level" (Newell 2001, 352). The critique of "modernity" and the state of "being modern" starts with a decentering of the singularity of the concept and its pluralization, thereby allowing *modernities in the plural*, or "alter/native modernities" (Appadurai 1996; Eisenstadt 2000; Gaonkar 2001; Holston 1999; Diouf 2000). What it means is the invention of modernities from within (hence native) and the capacity to alter incoming modernities so that they speak the native language and advance its identities. Witchcraft long deemed a relic of the primitive becomes a technology for negotiating and fashioning modernity (Geschiere 1997). In fact, the spiritual becomes the medium, scene, and force that tames, grounds, or refracts incoming technologies like mobile phones (Zegeye and Muponde 2012) that elsewhere constitute such state of the art as to configure the user (Woolgar 1991). African religious practice "over here" and cell phone design "over there" become connected and mutually shape each other (Fabian 1983).

This behavior of incoming things in local hands does not necessarily represent the far-reaching tentacles of globalization; in fact it also involves Africans themselves initiating the movements—of technology, capital, commodities, and other cultural goods. They are not necessarily appropriating modernities external to them, but are involved in a process of exchange, emitting their own things in exchange for those of the outside world. The goods are not just coming to them; they are actively constructing transnational networks through their own mobilities in the world—or those of their goods (Diouf 2000). Far from being a peculiar feature of today's cyber-connected world, this *extraversion* has been a persistent feature of African life, whether within the continent itself or beyond it, for millennia (Bayart 2000, 219). It is within this global engagement that Africa has provincialized or tamed not just the cell phone and, more recently, revolutionized its applications, but before it guns, bicycles, cars, and so on. This is not peculiar to Africans but to colonized subjects elsewhere as well (Chakrabarty 2000; Prakash 1999; Abraham 1998).

Perhaps it is not a question of incoming things (ideas, objects, order, or beliefs) taking over local modernities, but rather these incoming things as ingredients in the fashioning of "vernacular cosmopolitanisms" (Bhabha 1996). It is important, therefore, to account for "the process of globalisation and the multiplicity of individual temporalities and local rationalities that are inserted into it" (Diouf 2000, 680–681). In other words, incoming things become available or sought as instruments for being among others, for everyday utility, and for exercising being in the world, especially as prosthetic devices with which Africans help themselves to the global. This does not mean that they have to surrender their traditions and burn them before the incoming can enter without defilement. To the contrary, as Mamadou Diouf shows, Africans have time after time galvanized their constantly remodeled traditions and contrived their own economic scenarios to "anticipate a future saturated with projects of an indisputable modernity" (ibid., 683–684).

Approaches such as these could help us go beyond those of "Oh, that's terrible, these people are so helpless, let's go there and help them," toward those that say "Wow, these people are so creative. Could we be partners or clients?" The problem with reportage and images on certain parts of Africa mired in starvation or war is not that those images are fabrications, but the fact of their extension to all of Africa, at the exclusion of stories of happiness and creativity that are happening elsewhere. The consequence is that one image sanctions the NGO industry in the West to intervene among Africans whose agency or initiative is taken not to exist, whereas an image of Africa as innovative builds on what is already within Africa itself with Africans themselves as the central agent.

Technology in Africa, Africa in Technology

Technology fires the American imagination. Technology is "the American theology"—white America that is, and "scholars of technology often sidestep the subtext of whiteness within this mythos" (Dinerstein 2006, 570; Noble 1997). So pervasive is technology in American society that asking "What is technology?" could be a very stupid question. Perhaps even more stupid is "What is industry?" Only someone from a place without technology or industry like "Africa" (as if Africa is a country) can ask such a stupid question. Even to the earliest Africanists of European ancestry, Africa was preindustrial before European colonization (Marks and Atmore 1980; Marks and Rathbone 1982). Technology was an artifact of "The" Industrial Revolution, and industrial Africa a product of industrial Europe.

So what then constitutes technology in the African context? The meanings and practices of technology we have inherited in Africa are imperial. In fact, imperialism was the highest stage of capitalism (Lenin 1917); and Africa was at the receiving end. It was "the progress and power of industrial technology" that equipped Europe's domination and exploitation of Africa. That is why imperial historians of technology—who should not be confused with historians of Africa—say imperialism marks the "triumph of technology, not ideology" (Headrick 1981, 3–4). The promise of raw materials, markets, and investment openings fired technological innovation; technology itself supplied the means to secure them.

This was the motivation; imperial historians of technology like Daniel Headrick and Michael Adas were concerned with the means of achieving imperialism. With better ships Europeans could now travel far; with quinine they could stay alive while traveling; and with the telegraph and radio they could communicate while on the move. Indeed, machines became the "measure of men," and the capacity to dominate others now lay in the very process of technology design. The ship from which Western man now disembarked was not just "a marvel of design and workmanship" or a symbol of Western innovative supremacy but also "a spur to overseas expansion" (Adas 1989, 2; also Adas 2009).

It is important to emphasize that this was technology and imperialism as viewed from the North, a narrative in which the promise of technology on paper or by word of mouth did not translate peacefully into practice. As David Edgerton (2007) has cautioned in direct response to Headrick and Adas, the spectacle of technology and the acoustics of praise in its name elide its actual experience. The behavior of technology in the spaces of design and use "at home" do not always map readily onto foreign lands. Only through examining "technology-in-use" can we better account for its meanings and practices. As other scholars have shown, it is hazardous to limit the concept of technology narrowly to iconic artifacts without examining the larger social, political, economic, and cultural systems within which such artifacts are situated (Marx 2010).

It has been noted that decentering essentialist meanings of technology as "the application of scientific knowledge for practical purposes and machinery and equipment developed out of scientific knowledge" risks rendering the concept applicable to anything.[1] Postcolonial scholars also say that the concept of technology is still too Western-centric and is in danger of promoting commensurability.[2] In reply, it is equally true that the pedantic application of the concept carries serious ethical and epistemological consequences to those who are dismissed as untechnological when those who talk about technology—and engineering—include in their definition things in Western society that preceded the term and that look exactly like the things in non-Western society that are excluded. That is why symmetry is a prerequisite for exploring alternative definitions. To whom does a concept owe its power and meaning?

In the African context, technology had been indivisible from race and ideas about Africans as having less intelligence than whites long before the concept was in common use in the 1930s in the West. Movement and development (evolution), civilization and progress—these were the concepts that animated descriptions of the technological in imperial and colonial circles and indeed among scholars. The German philosopher and historian Georg Hegel (1837/2007, 99) described Africa as having "no movement or development to exhibit" and belonging to "the Unhistorical, Undeveloped Spirit [which was] still involved in the conditions of mere nature." In *Heart of Darkness*, Joseph Conrad's character Marlow, who transports ivory down the Congo River on a boat, captures well Western man's movement silhouetted against Africa's undeveloped spirit (Conrad 1902). As late as the 1960s, historian Hugh Trevor-Roper declared: "Perhaps in the future, there will be some African history to teach. But, at present there is none: there is only the history of the Europeans in Africa. The rest is darkness. . . ." He summarized African mobilities to "the unedifying gyrations of barbarous tribes in picturesque but irrelevant corners of the globe" (Philips 2006). That is all Africa meant to Trevor-Roper: an "unhistoric" place (Trevor-Roper 1969, 6). This attitude of treating Africa as an alterity, always in comparison to the West, provoked one of the most enduring critiques of this genre from the late Chinua Achebe (1978, 12): "For reasons which can certainly use close psychological inquiry, the West seems to suffer deep anxieties about the precariousness of its civilization and to have a need for constant reassurance by comparison with Africa. If Europe, advancing in civilization, could cast a backward glance periodically at Africa trapped in primordial barbarity it could say with faith and feeling: There go I but for the grace of God."

The first histories devoted specifically to technology in Africa—as opposed to African technology—did not stray far enough from Hegelian stereotypes. In a raft of essays in the 1960s culminating in his *Technology, Tradition and the State in Africa*, Jack Goody (1971) blamed this technological malice on the absence of two aids to mobility and mechanization. First, the horse and the plow, which put a huge onus on labor, reduced output per hectare, and relegated African production to subsistence. In other words, Goody emphasized mobility as a marker of technology; technology was mobility, and mobility was technology. Second, the lack of the wheel prevented Africa from accessing several kinds of "energy transfer machines," like gears, pulleys, and levers," which enabled complex textile and metal industries. Walter Rodney (1972) shifted the blame to another space: the export of Africa's human capital as slaves and its mineral and agricultural resources as industrial raw materials. Europe's technological development took place at the direct expense of Africa's underdevelopment. Technology had come down to incoming things and Africa's capacity to adopt or adapt to them.

But what did African attitudes to incoming things mean? For some it signified rationality, and those who didn't accept incoming things were irrational (Malowist 1966; Williams 1974). For others, such rejection was rational: the incoming things were not technological (that is, usable or useful) for performing the tasks Africans were doing anyway with what they had. Incoming things and their adoption were not prerequisites for industry or technology to occur. Africa was already doing well on that front on its own if one defined technology and industry in a relativist and contingent sense (Hopkins 1973; Curtin 1975). If one saw them from the viewpoint of a specific society and what its objectives were relative to the task and the tools necessary to execute it, such rejection of incoming things demonstrated that what was technological was Africans, not the thing coming into their domain. Technology did not come a priori; society never existed outside the technological, but rather, flourished because of means of living and being alive.

In writing a narrative of African creativeness, how then does one avoid ordering the story in such an "already appropriated" register? The approach I take is to begin with African technology and its itineraries, on the one hand, and incoming technology and its itineraries, on the other. This is an instance where fields like Science, Technology, and Society (STS) can help us understand where this incoming artifact is coming from and how its originators think they can delegate to a thing, by virtue of its interpretively flexible materiality, the power to configure users (Latour and Woolgar 1979; Akrich 1992).

Two sets of questions become interesting at the moment of encounter between this incoming technology and Africans. First, what happens to these properties when the incoming thing comes into contact with Africans? Does it come already as a technology and configure the Africans as users, or do Africans assign it meanings and functions as a *means (if that's what we mean by technology) of performing specific projects* of their own? That is what I mean by technology throughout the book. The second set of questions then becomes: What happens if we also extend the register of "designer" (or its equivalent) so that it is no longer just the scientist or engineer who invented the thing that then travels, but also the African who is coming into contact with, or importing, it? What if the "laboratory" is no longer the Western building where science is practiced, but the crop field, the forest, and other "open" and (en)closed places where knowledge is made and turned into tangible practical outcomes? What if we invert the subject of analysis, such that it is no longer just incoming things that are "interpretively (in)flexible," "(im)mutable mobiles," or "inscription devices," but also African technologies? The "African" here refers to what might otherwise be called "indigenous"—by which I specifically mean things derived from within and by African societies.

In other words, one hermeneutic and epistemological move might involve an act of symmetrical treatment of true and false statements, successful or failed science and technology, and human and nonhuman agents in the social studies of science and technology (Bloor 1976; Callon 1986; Latour 1987). In this particular case, the same concepts we use to analyze northern-made technology and science must be the same ones we extend to an analysis of African thought and practices. Unless we do so, we are likely to assume that the North is the domain of designers and the South of users, that "things northern" are technology and "things African" are primitive stuff that always give way to or contaminate "technology." Instead of being a mere user, the African becomes a designer who makes technology, not just someone who appropriates or (mis)uses incoming technology.

The point is that the meanings of things, not least the technologicality or technologyness of things, is contingent. As Gabrielle Hecht (2012, 15) found with the meanings and uses of the term "nuclear," there is no one ontology of technology but many: "*Nuclearity is not the same everywhere. . . . Nuclearity is not the same for everyone. . . . Nuclearity is not the same at all moments in time.*" Delete "nuclearity" and insert "technology" and the argument I make is the same, albeit from the trajectory of indigenous traditions. At stake here is what constitutes technology and in what context—where, to whom, and when? For the remainder of this book I explore technology from one site where the concept might tell us something we do not hear often in prevailing narratives—the site of ordinary people and their innovations or creativities, things that few would consider technological.

The major challenge remains one of not only the contingent meanings of technology, but also the trajectories: our choices of the subject of study and the starting points matter. Having set up need to decenter the North and the colonial library in narratives of Africa, even while accounting for its influences, I identify the challenge as twofold. First, to attempt a *longue durée* of Africa as a time traveler, already embarked on a journey, with Europe as a disruptive junction that reshapes the itinerary, slows it up for a time, then is cleared as an obstacle and the African journey resumes. The second is to see the African as a spatial traveler whose mobilities are not merely conveyances across geographic space but transient workspaces, with people as engaged in work-in-transit.

Defining Technology from a Transient Workspace

VaShona, who constitute the majority population of Zimbabwe, have the word *nzira*. It means a way (of doing things). As path, *nzira* is a venue where *kufamba* (movement) or *rwendo* (journey) takes place. The wayfarer (*mufambi*) is one who walks this *nzira*, in its specific meaning as a path or itinerary.

Kufamba is not only conveyance; it is also *kushanda* (working). The proverbs *chitsva chiri murutsoka* (something new is in the foot) and *mukuyu hauvinge shiri* (a fig tree does not come to birds) demonstrate the necessity, meaning, and function of traveling as work. This is what vaShona had in mind when coining the proverb *kukwira gomo husendemara* (to climb a mountain is to go round it, instead of directly and straight up). That is why *kashiri kasingapambare hakanune* (an unadventurous bird will never grow fat). The word *kupambara* defines clearly the idea and practice of mobility as creative work. It is a moment either deliberately contrived as a quest for opportunities, or simply an anticipative mode, calling attention to vigilance for the popping up of opportunities in the course of a journey intended for entirely different purposes.

These broader meanings of *nzira* as a site of creativity and work take us to a second, more collective meaning of "way": namely, *kufamba* (mobility) as a method of doing specific work, or *work through mobility*. Technology is a way of doing. The mastery of ways of doing certain things is called *unyanzvi* (expertise). As expertise, *unyanzvi* is a combination of *shavé* (spirit), *chipo* (talent pl. *zvipo*), *zvidobi* (skills), and *ruzivo* (knowledge). The success

Figure I.1
Nzira is commonly associated with the footpath through the forests, created by the shuffling feet of the *mufambi*, and, once established, becomes an infrastructure for navigating through the land.
Source: Black Bvekenyas Project (hereafter BBP) 2012

of any way and means of doing, let alone *unyanzvi*, depends on the guidance of the ancestors, who after death return in spirit to take care of the living. This is precisely why in times past, those who distinguished themselves in a specific *unyanzvi*—such as *mupfuri* (ironsmith), *hombarume* (professional hunter), *hurudza* (master farmer), *n'anga* (healer or diviner), *gwenyambira* (*mbira* player), *maridzangoma* (drummer), and *nyanduri* (master poet or court jester)—first consulted *vadzimu* (the spirits of their dead ancestors) to guide them in their practices. That is also why throughout the processes of metallurgy, hunting, farming, healing, musical performance, and oratory, the ancestors were asked to guide the proceedings, and those undertaking them observed specific codes of conduct (taboos).

MaHlengwe (the Hlengwe people) of maTsonga origin who are now called maTshangana also practice collective work—as do most other, if not all, African societies. *Hlengwe* means "gathering"; it exemplifies life as a collective action. It is said that when maHlengwe first arrived in

Gonarezhou from the coastal regions of southern Musambike (which the Portuguese corrupted into Moçambique), they would prioritize gathering first as a prelude to moving either together or in a coordinated way. Thus the land they settled, inhabited by vaShona, eventually became known as *Hlengweni*, the place of gathering, or where those who gather to work or partake of the fruits of work live. VaShona call such a gathering *kuita maonerapamwe chuma chemuzukuru* (the teamwork required for making the necklace of a grandson).

One specific type of collective work, moreover one that involves mobility, is hunting. It is here that vaShona and maTshangana epistemologies and practices of "the way" as both a path traveled and a means of doing things could tie together mobilities and provide new tools for excavating technology, telling African stories about innovation. *Nzira*, for example, would illuminate elements of transport, technological hardware and its commitment to means and infrastructures (Banister 2005). It would extend our understanding of mobility beyond familiar Western(-originated) high-tech exemplars like mobile phones, cars, trains, airplanes, airports, tourism, and the city (Adey 2009; Urry 2000; Sheller and Urry 2006; Cwerner, Kesselring, and Urry 2008). It would find a home among scholars seeking to understand technologies of transport and their historical dimensions (Reuss 2008) by broaching the subject beyond automobility, trains, tourism history, and sociotechnical structures that animate Western discourses.

Mobility matters to this study only as a methodology for exposing technologies of everyday innovation and the productive value and role of movement. This is the point at which the book enters into dialogue with both transport history and the new mobility paradigm, which are still nonetheless framed with Western- and artifact-centered grammars. Indeed, both the transport and mobility discussions could benefit from an African perspective and other vehicles and mobile objects besides cars, cell phones, internet, airplanes, or trains. It should still be possible to have a conversation on mobility even after "technology" has been removed completely from the equation. Even though some working within transport history think "no turn is necessary" and wave away their critics to "turn if they want to" (Freeman 2006), those that see this need (Mom 2004; Divall and Revill 2005, 2006; Walton 2006) have a far better chance of engaging in conversation with a wider audience. The reason is that transport alone is too limited to account for *the work of moving*.

The strength of *nzira* lies in its double placement of the path and its traveler (way) and the ways and means with which the task is accomplished in one narrative. The first formulation of *nzira* places the traveler in a *transient place*—that fleeting moment when body and place meet, when bodies meet in place (physically or virtually) (Jirón 2010, 68). The second equips the itinerant with the means to transform the transient place he occupies into a transient workspace. Here, Karl Marx's concept of *labor-power* as "the aggregate of those mental and physical capacities existing in a human being, which he exercises whenever he produces a use-value of any description," becomes important. Labor-power, Marx continued, can be realized "only by its exercise; it sets itself in action only by working" (Marx 1867/1954, chapter 6). Marx was careful to draw a sharp distinction between *labor-power* (ability to work—muscle power, dexterity, brain power) and *labor* (the actual activity or effort of producing goods or services, viz. use-values).

In its expression of *nzira*, the hunt of vaShona and maTshangana is an example of a *transient workspace—an area, site, or space where mechanical work is being performed as and because the body is moving*. Thus mobility ceases to be just conveyance from one point or state to another, but production-on-the-move, as opposed to Marx's view of a workshop (factory) as being tethered to immovable places. The hunt is a perfect site to observe incoming things and Africans encountering each other. Instead of claiming what each did to the other from a geographically fixed place, and missing the microparticulars that characterize encounter, the hunt is located on one type of *nzira*: the spoor. Upon following the hunters that are following the tracks to arrive at and kill the animal that made them, a history of African technologies—and not just technology in Africa—is revealed.

An Outline of the Book

Transient Workspaces is divided into eight chapters. The first explores vaShona and maTshangana's philosophies of all human mobilities as guided by ancestral spirits (*vadzimu*) and treats the forest as a sacred space. To be able to navigate the forest required a specific understanding of the spiritual relationship among *Mwari/Xikwembu* (God), ancestral spirits, the living, the animal world, and indeed the trees, rivers, and mountains. As guided mobility, the hunt poses interesting questions on what constitutes technology under regimes of spirituality. The site of analysis for this encounter between African innovations and incoming technologies is the professoriate of the hunt, defined in chapter 2 as a spiritually guided institution and practice that educated boys in the chase through doing. In chapter 3, the professoriate is examined as a mechanism through which maTshangana and vaShona interpreted incoming things like the gun, to which they assigned new meaning and uses, even as they acquired new competences through interactions with it.

The next two chapters are dedicated to how European colonizers under siege from the deadly tsetse fly deferred to these hunters in the absence of any remedies of their own. Chapter 4 sets the scene by exploring the disruptive role of European colonial partition, mobilities, and settlements. This "ecological imperialism" (Crosby 1993) involved Europeans displacing Africans from lands they had tamed, in the first instance, and settling in areas where Africans had sequestered pestilent insects like tsetse fly, in the second. Then the pests struck back. In the absence of any ready biomedical or chemical solutions, the colonial state turned to the African hunter to help slaughter forest animals, starve the tsetse fly of its food source (blood), and kill the deadly trypanosome protozoan the insect transmitted from forest animal reservoirs to livestock. This is the subject of chapter 5.

The last three chapters turn to the criminalization of African hunting in the wake of emerging wildlife conservation regimes during the colonial period and the uncritical retention of the same under postcolonial conditions. The fact that "fortress conservation" has not delivered security against poaching, and has only served to criminalize what could be a powerful ally and player in wildlife sustainability, calls for a rethinking of approach. However, the hunt is one of many criminalized knowledges and practices to which people defer in times of crises, particularly the crisis of "modernity." Since most of these approaches were not informed by or for the benefit of Africans, and since the colonial regimes that installed them are gone, it is important to critically revisit them with a view to constructive engagement.

1   Guided Mobility

In the way of life of vaShona, all mobility was guided mobility. *Vadzimu* (ancestral spirits) were the guides, who interceded for the people to *Mwari* (God). *Vadzimu* had many hierarchies. From the most senior, there were the *mhondoro dzematenga* (spirits of the skies or atmosphere), then the *mhondoro dzedzinza* (clan spirits), and, most junior, *vadzimu* (family spirits). The heavenly spirits were the gatekeepers over rain and all meteorological matters. To them the elders appealed in times of drought, asking these ancestors to pamper the land with their saliva (*matê*), their tears of mercy, to pull the clouds directly above them, that the rain might fall on top of their famished crops (Mukute 1983a, 65).

As hydrological mobility, precipitation—raindrops pattering from the skies—provided the moisture the fertile wife of Mwari, earth, needed to bear children (crops), to become a mother. Without it, earth was *mhandje* (a barren woman). *Mhondoro dzematenga* were sources of unlimited power and knowledge, including as seers or prophets (*vafemberi*). The most celebrated of these *vafemberi* in Zimbabwean history is Chaminuka, who predicted the coming of *vanhu vasina mabvi* (men without knees, or who wore trousers) and *hondo yechindunduma* (war of the booming cannon), the anticolonial resistance otherwise known as *chimurenga*, in 1896–1897. *Mhondoro dzedzinza* asked for rain on behalf of their clans, and could be elevated into *mhondoro dzematenga*; at that point they catered for the entire *nyika*, not just their own *dzinza* (Nyevera 1983, 24–5). To help them on their way in this onerous task, the living offered them *fodya yegonamombe* (tobacco for the smoking pipe)—which was *mbanje* (marijuana) for the *mukombwe* (medium of a rainmaking spirit)—so that the *mhondoro* might travel in a good mood on the journey to ask for rain (Mukute 1983a, 64).

A strong symbiosis existed between vaShona and other creations (*zvisikwa*, or the environment at large). *Vadzimu* (ancestral spirits), *vanhu* (people), *miti* (trees), *mhuka* (animals), *makomo* (hills), *nzizi* (rivers), *mweya* (air), and *nyika* (earth) were all creations of *Mwari* or *Musikavanhu* (creator). The visible and the invisible were hidebound together. Everything that was *panyika* (on earth) was also replicated *kunyikadzimu* (where spirits lived) (Nyevera 1983, 20–21). The earth as the wife of Mwari was the mother of humankind, hence 'motherland' whose breasts the living suckled for milk (water) (Majoto 1983, 44).

Every facet of life was spiritual—*kuzvarwa* (birth), *kuyaruka* (growing up), *kuroora* (marriage), *uhurudza* (farming prowess), *ushe* (chieftainship), *utongi* (governance), *urwere* (illness), *kufa* (death), *kuvigwa* (burial), *kudzoswa mumusha* (a dead kin's spirit being brought back into the village), and *kubatwa zita* (being divined as a future spirit medium). Life was a journey that the living entrusted (*kukumikidza*) to *vadzimu*; the ancestors were the sight, the eyes, of the living, without them the mortals were *mapofu* (blind people) (Mukute 1983b, 109). To grow, a clan needed fertile wives. *Vadzimu* were asked to guide the men in the direction where fertile women who were not witches (*varoyi*), gossipers (*vanamuzvinaguhwa*), sharp-tongued (*vano-chenama*), and lazy (*vanamanungo*) were to be found (Nyevera 1983, 19).

Death was the process not of final expiry but merely an elevation into an ancestral spirit. Dying was sleeping (*kurara*); it was a stage in the process of being called to spiritland (*nyikadzimu*) (Mashiringwane 1983b, 142). Burial was laying to rest (*kuradza*) or to hide (*kuviga*). The chiefs were usually interred in the armpits of the hills, their mummified bodies stood against the cave walls, along with their shields, bows and arrows, and spears, that they may continue hunting in the forests of *nyikadzimu* (Shumba 1983c, 152). People were usually known by the name of their chief, hence *vanhu vaIshe Chitsa* (Chief Chitsa's people), or their totem, hence *vayeraSoko* (those who taboo monkeys). Places too were known, among other ways, according to the chief who ruled them, hence *dunhu raShe Chipezeze* (the land of Chief Chipezeze), or the totems of people who lived there, for example, *kuvayera-Nondo* (the lands of those who taboo the tsessebe). These references were not just homages to the living mortals, but also to the living departed, who were resting in the hills and the dark forests.

The funeral was not merely a final goodbye with the dead; mourning—the quality and quantity of crying—demonstrated the deceased's impor-tance to the living; the family had to "love" the corpse (just as Hartman [2007, 70] describes). The funeral was not just a vehicle for spirits to travel into spirit worlds; the properly buried person returned as an ancestral spirit, the badly buried or unburied a ghost and a vengeful spirit.

The funeral being over vaShona waited until a summer had passed before returning the dead into the village as a spirit. After "sleeping" for a while the dead was awakened to look after the family. This was called *kurova guva* (beating the grave) or returning the deceased back into the homestead or village (*kudzora muchakabvu mumusha*). Between the death and the *kurova guva*, the deceased's spirit would be wandering in the forests like *dzangaradzimu* (ghost) as if she did not bear children (*kumbeya-mbeya nemasango kunge asina kubereka*) (Shumba 1983a, 137). At *kurova guva*—involving *hwahwa hweguva* (beer for beating the grave)—traditional beer was sprayed around the grave, while an elder called the deceased by name. This ritual was conducted in the time of *mashambanzou* (elephant bath time, in the period before dawn began to break), which was also the time when *varoyi* were returning home from their nocturnal itineraries (Shumba 1983c, 152). The deceased was told that the days of roaming the forests were over; the family had come to take him back into the freedom of his own homestead, where he could resume the duties of looking after his children again and protecting them against vulnerabilities (Nyevera 1983, 20). The process of returning the dead home, allowing the spirit to enter the village from its wanderings in the forests outside, was one of collective rejoicing. At last the tears shed at the death and burial could now be wiped away (Mahachi 1983, 120).

A life and worldview so thoroughly spiritual invites a number of questions. First, what becomes of the concept of *nzira*—or the definitions of technology and mobility—in a situation where all movements, including even of animals and within the forests, are so thoroughly dominated by the spiritual? What becomes of expertise, of skill, and indeed the materiality of things? Such an inquiry may, for example, give a perspective of technology that escapes the severely limited and even absent consideration of the spiritual in Western and most Africanist studies of technology. What follows is a discussion on the limitations of imported concepts and the benefits of finding alternative epistemologies, in the first instance, and overcoming Africanist reliance on the colonial library by searching for archives that Africans themselves have produced. Arguing against reducing African philosophies to archives, in the rest of the chapter I then elaborate on the epistemology of guided mobility (*kutungamirirwa*) and what it says about technology under sacred conditions.

After Imported Epistemologies and Colonial Libraries of Technology

This chapter addresses technology from the perspective of vaShona philosophy of life as spiritually guided mobility. Using Shona praise poetry (prayers), proverbs, and registers, it proposes the concept of guided mobility to examine what technology might mean in a realm of ancestral spiritual sovereignty over the living. Methodologically, the chapter invests philosophical and epistemological value in Shona knowledge to trace the origins of ideas and practices that European colonists later appropriated as their own. This is a rather unorthodox narrative of African history, technology, and mobility because it starts from African knowledge as philosophy, whereas others treat such knowledge as an archive and Africans as informants, and conceptualize their work basing on Western theories.

The Shona concept of *nzira* could address the dual dilemma that mobility studies are struggling with at the moment. On the one hand transport history is rather limited to rigidly Western-centric modes of transport (cars, railroad, trains, roads, etc.) or phenomena easily legible as "technology," thus ruling out a vast array of other "non-technological transport." On the other, the "new mobilities paradigm" does not have much, if any, historical purchase. Both are still rather banal mobilities, that for the non-Western reader, explain how Western technology and how people in the West and elsewhere have "responded" to it without ever leaving the comfort zone that Western epistemology has built for itself in terms of understanding the world from elsewhere. By 2009, the more flexible elements in transport history were pushing strongly for a platform that publishes work that is not conservative enough for the *Journal of Transport History* (with its auto-centric and technocentric emphases) and too banal for the more theoretical *Mobilities*.

I met one of these scholars, Gijs Mom, in 2009 when he was discussing the possibilities of launching that sort of journal, which eventually became *Transfers* (published by Berghahn). I was coming from neither tradition; in fact, my insistence and critique, even of *Transfers*, was and has remained that it cannot just aspire to write "an accessible, theoretically assured, richly descriptive history of transport-cum-mobility grounded in the material realities of technological systems and alive to issues of power, social equity, and ecological sustainability" (Divall 2010, 950) and end there. Transport is not the sum total of mobility, nor is all transport necessarily technological (Mom 2004, 121–122; Walton 2006). It is not enough to simply free up transport history to embrace culture, or even the theories of new mobility studies, and move on. As long as the subject of discussion remains transport- and auto-centered, a whole slew of other meanings and functions of mobility besides conveyance are foreclosed. In the African context, transport history has evolved around auto-, rail-, and aero-centered narratives (Pirie 2011, 2009a,b,c, 2008, 2006, 2004, 2003, 1997, 1993a,b,c, 1992, 1990, 1987, 1986, 1982; Monson 2009; Gewald, Luning, and van Walraven

2009). Beyond these banal mobilities, there are multiple other movement-centered studies of Africa that Mavhunga (2012) surveys.

While the transport history debate rages on, the new mobilities paradigm has emerged along three main trajectories of enquiry. The first relates to *mobility-subjects* (individuals and groups using or affected by transport systems). The second relates to *mobility-objects* (motor vehicles and their infrastructures conceived as "hard" and "soft," namely, roads and social-institutional, respectively). The third concerns *mobility-scapes* (the shaping, perception, representation, and performance of time and space). In proposing a "new mobilities" paradigm, John Urry and Mimi Sheller were thinking of "diverse mobilities of peoples, objects, images, information, and wastes; and of the complex interdependencies between, and social consequences of, such diverse mobilities" as well as transport history, with its commitment to historical context (Urry 2000, 185).

I share this commitment to "diverse" mobilities while noting the need to bring in the historicity of far broader meanings and practices beyond the familiar ones—cars, cell phones, airplanes, airports, trains, tourists, and passengers. African idioms to which I defer offer not only different empirical-historical material, but also their own African *Weltanschauungen* that are thinkable and "explicit within the framework of their own rationality" without necessarily being overburdened with the transport history and new mobilities debates (Mudimbe 1988, x). For however original the new mobilities paradigm might be, it represents specific Western orders of knowledge and their specific histories. They betray a need to understand contradictions and emerging puzzles within Western conceptions and practices of mobility. They will not allow the study of mobility as *nzira*, to be understood endogenously, at its own terms, and deep within its own belly.

Deep Shona, Deep Idioms of Mobility

Just as Western scholars of mobility might defer to the new mobilities paradigm for theory, I defer to African proverbs, praise poetry, and other idioms as philosophy and practices, and the elder-led institutions of education and apprenticeship as professoriates. I am an insider to these idioms; I was apprenticed in this professoriate of indigenous knowledge through the guided mobilities of boyhood and young adulthood. The written texts to which I defer when necessary are written in the Shona language I speak.

Therefore, I have "the credentials to produce an African epistemology [of technology and of mobility], describe it, comment upon it, or at least present opinions about it" (Mudimbe 1988, x). There is "deep Shona," which only insiders and trusted and immersed outsiders can understand, which is seldom accessible to one with a few months or years of language classes and "participant observation." The latter can only acquire competence in ordinary Shona; at length, when they have "gone native," when they can be trusted, then the deeper knowledge is revealed to them.

The "deep Shona" texts that appear in written form are translations of the first and second generations of African scholars and immersed Europeans who believed that vaShona had a philosophy of life worthy of documentation and understanding from deep within the caverns of its own rationalities and rationales. These scholars first preserved its spoken word into written Shona, trying their best to translate the acoustic character and accompaniments using phonetics. If something was lost, perhaps it is the smell and the visual. Any theory that exists in this praise poetry, proverbs, games, and registers derived from the African ancestors themselves, who had handed them to the living while they themselves still lived in flesh. Nor does one need to worry that the intellectuals translating these idioms into written text might not have understood the subject of their writings (Mudimbe 1988, xi). They were the subjects of the writings they were translating and lived – were embodiments and pronouncements – of these texts' lived lives and realities.

This is knowledge that was produced by, in, and for the collective; as poems, registers, or games, the person who put them to paper is not necessarily the author but rather a compiler of knowledge collectively authored, or knowledge-made-in-common. Nobody can claim individual title to knowledge that the entire community produces. In the Western tradition, only the individual "thinks and constructs knowledge," even if that is done within and for an institution like the university. Hence individuals construct *knowledge* (facts) while the rest of society is out there constructing *myths* (Masolo 1991, 999).

Yet, even the collective is incomplete without the ancestral spirits, the bedrock upon which the inspiration to innovate is built. This begs the question: What happens to skill or expertise and to technology in the realm of the sacred?

What Is Mobility in the Realm of the Sacred?

The dead could be consulted at sacred places, trees, forests, mountains, springs, and pools that never dried up. Only those undefiled, such as virgins and elderly women, could approach them (Nyevera 1983, 36). The sacred were no places of play or loose talk, otherwise the forest would feel slighted. Pointing at things or laughing at them also carried grave consequences. Such actions were considered to be cursing and slighting *Mwari*, creator of everything in the forest. Upon meeting an animal, especially one's totem, if one did not clap hands (*kuombera*) in respect, the traveler was forbidden from passing disparaging remarks. This would be considered an insult to *mhondoro*, the king of the forest and great ancestor of the nation. It was tantamount to "*kukanda mhiripiri mumeso avadzimu*" (throwing hot peppers into the ancestors' eyes) (Nyevera 1983, 18).

Forests were full of wandering spirits that could easily get antagonized or cling onto people of loose talk or those who did not respect protocols. These spirits inhabited *matongo* (former village sites) where people died and were buried, their descendants subsequently relocating elsewhere (Hodza 1979, 237). The forests were teeming with *madzimudzangara* (wandering ghosts of people who died), roaming in thickets, at pools, in mountains, and around burial places at night (ibid., 206). They were spirits of foreigners like Portuguese traders and Nguni warriors who died and were not given proper burial (ibid., 353; Bullock 1950). Some were simply spirits of baboons, pythons, and other animal species, what one commentator called "infra-humans" (Hodza 1979, 353).

The forest was also where bad spirits, *urwere* (illness), and those struck by dangerous illnesses like *maperembudzi* (leprosy) were cast away, left to wander (*kurasirirwa*) or to die. *Ngozi* (vengeful spirits) were left in the forests (cf. Silla 1998, 35, 73–75). The spirit was first cast onto a black goat or chicken with the pronouncement: "Do not come back because *you have been given legs to walk with*, hopefully your ungrateful type will not . . . return tomorrow." The goat was then left in the middle of *sango*, so that it could not retrace its steps. It then wandered, carrying its bad spirit, unable to find the way back to the village. Anyone who came across it and thought they had found "wild fowl" would become the new organic vehicle of the bad spirit (Shumba 1983, 352).

Forests were sites of prayer and sacrifice, conducted at or under specific trees, especially the *muhacha* (cork tree) (Hodza 1979, 246). A *nhandare* or *pamutoro* (sacred enclosure) was constructed surrounding the giant tree, which was regarded as the *mhondoro's* home or house (*musha*). Here, the living could talk to the spirits, with beer and poetry as devices for the transmission of their yearnings. Newly harvested crops were placed under the *muhacha* and presented to thank the ancestors for rain; trees, therefore, were not just plants but also communication infrastructures with the ancestors—and *Mwari*. It was therefore fitting that the spirits be first to taste the crops before the children had eaten. Every household selected only the best

Figure 1.1
Mhondoro, the lion, occupies a dual (spirit-animal) realm as lion spirit and king of the forest.
Source: TKAV, 2011

and biggest crops, including *zviyo* to be used to brew beer that *masvikiro* (spirit mediums) would drink (Nyevera 1983, 24).

It is worth recalling that *mhondoro* also means "lion," king of the forest; it also denoted the ancestral or clan spirit. *Mhondoro* manifested itself and spoke to the living via its human medium, the *svikiro* (port of arrival), growling like a lion before and between words. The *mhondoro* intersected many life-worlds; it connected the burial of flesh, the birth of an ancestral spirit, and the lion king. VaUngwe people—a Shona-speaking community—collected the liquid drained from the deceased chief's mummifying or decomposing body and buried it separately from the corpse. Out of this liquid was believed to rise the *mhondoro*, hence people were forbidden from killing a lion lest it be the reincarnation of the chief. The Saunyama—also vaShona—people of Nyanga believed that a lion cub "takes up its abode in the grave and there it is fed by guardians of the tomb" (Posselt 1935, 82). When sacrificing a black bull or ox to the *mhondoro*, some vaShona societies left it in the open overnight. If they found the carcass gone, the spirits had accepted the sacrifice; if not even any footprints were found, then they had rejected it and were not happy about something.

Being therefore a place where spirits lived—in trees, among the rocks, and in animals—the forest was a space requiring ancestral guidance for anyone moving within and through it. Among the *vayeraTembo* (those who tabooed the zebra), anyone passing an ancestor's grave was expected to place upon it leaves from the *mupfuti* tree to honor the dead. The tree's fiber was used to tie roofing, rafters, and withies together; so too the ancestor, who tied the nation—indeed life itself—together and gave the living a firm grip on their location among others and on the land, that they might navigate purposefully and fruitfully.

On a long journey through the dark forests, ancestral guidance was recognized and experienced in many ways. It could come in the form of a bird. The sight of *chapungu* (bateleur eagle) represented the reassuring presence of *vasekuru* (grandfather), the ancestral patriarch. When *chapungu* circled above the home of a person struck by illness, recovery was assured; the ancestors would have refused to accept his spirit into *nyikadzimu* yet (Nyevera 1983, 26, 29). In this sacred bird's mobilities and acoustics was a message—the ancestors sent it especially when the traveler was walking on the road or camping in the forest, or in the heat of battle, to warn of impending danger or to assure that everything was well (Daneel 1995, 84–120). *Chapungu* communicated through its wing flaps, somersaults, its distinct *Kovo-o-o* shriek, or its silent, peaceful flight. The ancestors might send *kamba* the tortoise (or turtle) and *kovo* the squirrel as well. If *kamba* kept walking and the squirrel crossed the path ahead of a traveler in a tail-down position, the way ahead was good; if the *kamba* stopped and *kovo* crossed "tails-up," danger lurked ahead and the journey was to be terminated forthwith (ibid., 15).

What Then Is Technology in the Realm of the Sacred?

Vamazvikokota, amai vebere is "deep Shona" for "expert, mother of a hyena" (Hodza 1979, 277). The term *amai vebere* (mother of a hyena) refers to a woman of talent, and stems from the way young hyenas go everywhere with their mother, who weans them only after their instruction is complete and they are free to take care of themselves (ibid., 336). It embodies all elements of expertise, as does *shangura*, another word used to describe an expert at specific work or craft—in building (men), ceramics (women), a gifted speaker, a prolific hunter, a smith, and so on (ibid., 355). Such a person could only be likened to *n'anga rucheche* (diviner infant), deep Shona for a distinguished diviner who did not miss, who revealed the "truth" always.

The term *shavé* translates to "genius" in English. It was a spirit (*mweya*), but unlike *midzimu*, *shavé* was the spirit of a deceased *mutorwa* (foreigner) manifesting in a clan member. When possessed, the *svikiro* spoke in a foreign language he or she did not speak when not. Good *shavé* or *rombo* could be a guardian spirit that "accompan[ied] the person through life and [gave] direction to his interests, powers, and abilities" (Hodza 1979, 27). With its guidance, a person could choose to pursue—and became successful at—such pursuits as divining, hunting, metallurgy, farming, instrument playing, dancing, and so on. It was the perfection at one's work that led people to exclaim: *"Hazvisi zvega [izvi], pane chinokwenyera"* (This is not just skill on its own; there is more to it) (Nyevera 1983, 27; Hodza 1979, 299). *Shavé* was active interest nurtured through doing, trial and error, and discovering. It did not come to slaggards, sloths, and fools who just "sat there!" and waited for fate to take its course.

However, a *shavé* got offended when the individual boasted or claimed undue *credit* for the skill and prowess. After all, the presence and agency of *shavé* within the hunter, smith, farmer, or fisherman would have been acknowledged through *hwahwa hwamashavé* (beer of the *mashavé*) as behind the individual's exploits. It was the *shavé* that protected the individual from harm and gave him good luck (Hodza 1979, 353).

Mashavé was a combination of spirit (*mweya*), talent (*chipo*, pl. *zvipo*), and an individual skill (*unyanzvi*). *Shavé* involved skilled work; and as already indicated, all work among vaShona involved mobility of some sort. The *shavé* was witnessed to have possessed a person when he was able to do specific things exceptionally well. The person also possessed *shavé* through the talent *vadzimu* gave him and the skills he acquired through training. This convergence of capacities made certain persons into *mupfuri* (ironsmith), *hombarume* (professional hunter), *hurudza* (master farmer), *n'anga* (healer or diviner), *gwenyambira* (mbira player), *maridzangoma* (drummer), and *nyanduri* (master poet or orator). These tutelary *mashavé* were distinct from the bad *mashavé amapfeni* (spirits of baboons), *mashavé amazenda* (foreigners killed in war and never buried properly), or *mashavé amajukwa* (water spirits), all manifesting through mediums and making demands of gifts in the expectation of appeasement. They were also distinct from bad spirits that elevated skills like *kuba* (stealing), *kukara* (gluttony), *mangoromera* (boxing), and *upombwe* (prostitution). These bad spirits were cast out of a person and into the forests whence they came (Nyevera 1983, 27; Hodza 1983, 353).

Figure 1.2
Mhizha, the smith, from whence all the community's metalware used to come.
Source: The Duggan-Cronin Collection

Mhizha: The *Shavé* and Work of Metallurgy

Mhizha was any craftsman skilled in the arts of smithing, weaving, carving, and the building of houses, granaries, and other structures. *Nyanzvi* referred to skill in these arts and those of dancing and the playing of musical instruments. Special titles of *maridzangoma* and *maridzambira* applied to the drummer and *mbira* player respectively (Hodza 1979, 25). As a smith, *mhizha* was also called *mutare*, the one who smelted ore (*mhangura*) into iron and metals into finished artifacts (ibid., 352).

The presence of iron and the importance of tools for agriculture, hunting, and trade made the ironsmith a very important member of Shona society. Among many other sources, the Zezuru-speaking people obtained ore principally from three mountain ranges: Hwedza, Biri, and *Murwi wemasimbe* (The heap of charcoal) near Shamba (Shamva). Individuals or groups of ironsmiths made their own arrangements for *kuchera* (mining), *kunanautsa* (smelting), and *kupfura* (beating, meaning smithing). Each ironworker's "helpers" included *vazukuru* (grandchildren and nephews) and *vakuwasha* (sons- and brothers-in-law), and *madzimai* (wives and daughters) for portage of ore in baskets from mine site to smithy. Upon arrival, *vazukuru* and *vakuwasha* assisted the *mhizha* in blowing the bellows (*mvuto*, from the word *kuvhuta*/blowing) and other manual tasks. To become *mhizha*, an apprentice not only did the smelting himself, but also learned "the medicines and herbs needed to enable him to practice the craft without the risk of blindness." These *mishonga* (medicines) were only divulged to the person the *mhizha* had identified to take over his trade (Hodza 1979, 351; Editor [NADA] 1926, 25; Thompson 1949; Hatton 1967).

The meanings and social place of metallurgy among vaShona are captured well in excerpts from the poem (recorded in Kambasha Village, Musana, in 1965) "*Kurumbidzwa kwemupfuri wemhangura*," in which a wife pays homage to her husband's *shavé rekupfura* (spirit enabling him to smith). She implores those without hoes and axes to come and buy. Her husband, she says, is a "*shangura pautare. . . . N'angarucheche yokupfura mapadza*" (a craftsman in iron, truly a wizard at forging hoes). She adds: "*Murume wangu ishuramatongo pakupfura, mhizha inenge namo*" (My husband is an expert in working iron, a craftsman who sticks like wax to his trade) (Hodza 1979, 352, 355). When his spirit possessed him that day, the forge would be worked so hard that it would almost speak, the slag piling up into a mountain. The implements made spoke for themselves: piles and piles of *madimuro* (choppers), *makano* (battle-axes), *mbadzo*, *tsombo*, and *huhwa* (large and small hatchets), *serima* (small hoe for weeding millet), *kakuruwo* (a very small hoe used for weeding between finger millet stalks), *chifengu* (a large hoe for ridging), *mapanga emapakatwa* (double-edged knives), and *mbezo* (adzes) (ibid., 351–352, 354).

An English translation of a typical transaction between *mhizha* and his customers comes from a passage that a Shona oralist named S. Taoneyi compiled early in the twentieth century:

When the smith has finished forging all these things which I have mentioned, do you think he distributes them among all the people? No, he does not do that; people come to buy them with cattle, goats, sheep, and fowls. He who pays a beast is given ten hoes; who pays a goat, sheep, and pig, is given two; who pays a fowl, just one. That one who is without anything to make a purchase with, he just works for them having been given work by the smith and the days of working are spent. When these days are over, he is then also given his hoes, and he goes off. Some used to pledge one another a contract of friendship whereby they could get hoes, cooking for one another large brewing of beer. Another who is very much in need of something to use, takes his daughter and decorates her with beads on the head and anoints her, then he leads her in front and goes and gives her to the smith, saying:

"Look, I have given you a wife, you now keep on forging for me on your side hoes, choppers, axes, token axes, and everything else." The smith is grateful, and he also gives everything to the father-in-law. (Hodza 1979, 354)

Hurudza: Community as Labor Power

The *shavé* of a *hurudza* manifested in the abundance of agricultural possessions, exhibitions of abundance, and polygamy. His cattle herds produced *nyama* (meat), *mafuta* (lard), *mukaka* (milk), *sadza* (thick porridge), *matehwe* (skins), *nyanga* (horns for trumpets and making *makona*, or medicine horns), *ndove* (dung for decorative wall plastering and floor covering), and *miswe* (tails for making headgear, flywhisks, and part of a diviner's toolkit). Clan traditions say that my own grandfather, Mavhunga Shayirekudya, and his brother Nyembe Shayirekudya, would go out to look for beautiful women riding bullocks (steers) they would have tamed themselves. They deliberately smeared their knees with cow dung as a signature of cattle wealth, carrying *maguchu* (calabashes) full of liquid honey. Upon arrival at a well where nubile princesses were fetching water, they would ask for a drink. They would then feign to taste the content of the *mikombe* (ladles) extended to them, then give them back in haste, saying: "Why do you drink such *mavave* (sourness)? Here, try the water we drink where we come from." The combination of honey and cow dung promised a world full of good life. The charm worked; one of the women, Manungo Tsvande, would become my grandfather's wife, mother to my father, and the grandmother who departed for *nyikadzimu* too early for me to see. Like vaZezuru, vaDuma were *hurudza* whose cattle pens overflowed with cattle, milk, lard, meat, skins to manufacture *ngoma* (drums), cow dung to polish the floors (and keep away dust), and ox-tails for making flywhisks (Hodza 1979, 289).

Because it was a place of much food and wealth, *kuroorwa pahurudza* (being married into a master-farmer's household) was attractive to many women and fathers with daughters. In times of famine a *hurudza* son-in-law was a good place to go *kunosunza* (seek) grain; when marrying off and losing a daughter to him, a father expected many fat cattle to be driven in the opposite direction. The *hurudza* also knew that being a successful farmer needed many hands; besides affection and conjugal comforts, wives existed to bear many sons and daughters to work the land, to enable parents to receive *roora* (dowry) of their own in cattle when they were given away in marriage. The fertile soils, well-chosen when migrating, required

Figure 1.3
Many cattle, one of the markers of a *hurudza*.
Source: TKAV, 2011

the fertile wombs to produce the many hands necessary to work them into big harvests.

Occasionally *hurudza* instructed his wives to brew beer, then invited his best friends so as to show off his kindness and mollify those who harbored intentions of witchcraft against him. This kind of beer was called *hwahwa hwechinweinwe* (beer for merriment and relaxation), as distinct from *hwahwa hwendari* (beer for sale) and *hwahwa hwejakwara* (beer for collective work) (Hodza 1979, 246).

Also called *nhimbe* (chiShona) and *humwe* (chiTshangana), *jakwara* or *nhimbe* (work party) was a day set aside for the collective work of planting, weeding, and harvesting the fields of the orphaned (*nherera*), the widowed (*shirikadzi*), the elderly (*chembere* and *harahwa*, female and male respectively), the poor (*varombo*), and the master farmer (*hurudza*). Because *chara chimwe hachitswanye inda* (one finger cannot crush lice), it made sense to mobilize the community to create a large labor force.

The beneficiary of this collective labor brewed beer and hosted a *nhimbe* to provide not only an incentive for people to come but also the energy

Figure 1.4
Even today, beer brewing remains the domain and expertise of women among vaShona and maTshangana.
Source: TKAV, 2011

to continue working. Beer and food aplenty were served; after all, vaShona believed *ukama igasva hunozadziswa nekudya* (friendship is an empty vessel, it can only be filled up through a meal). Food and material possessions did not substitute for collective belonging; and food and material possessions were contents of a vessel called *ukama* (community). That is why *chivanhu* (culture) said that *vana vanyamunhu* (children of the same ancestor) must eat, drink, work, and pray together.

The collective was the water; only within it could the fish (the individual) find strength, hence *simba rehove riri mumvura* (the fish's strength is in the water). Then, like one organism, one body with many different but mutually coordinating parts, the village descended on the field as one, each bringing their own *mapadza* (hoes) ready to weed and *matemo* (axes) ready to chop. The individual increased his or her mechanical means from a pair of hands to tens and hundreds of hands, making the task light.

Singing and song lubricated difficult work into fun and sport, pacing it, giving it rhythm, and taking the mind away from the arduous task, and placing it in a parallel universe of recreation and entertainment, thereby shutting it from communicating "evil thought" of *nungo* (laziness), *kuzeza* (dreading the task), and *kuneta* (tiredness). Singing transformed and coordinated this critical mass of individuals and their limbs into one vast organism marching in lockstep. The food provided the fuel for the body and mind to remain energized and to perform the work at hand.

The release of beer to the workers was paced and dispensed in quantities designed to give them *simba* (strength) without driving them into *kudhakwa* (drunken stupor) that neutralized efficiency. The volume of the singing picked up, powerful arms swung the hoe this way and that. Soon work became just another excuse, another occasion, to drink, eat, and party. Work became a medium of entertainment. Soon the work itself was finished, and there was only beer, food, and singing to be partaken. With the hand, feet, and body now freed from the "small matter" of hoeing, *ngoma* (drums), *mbira* (hand pianos), *hosho* (percussion gourds), *magagada* (ankle percussion gourds), and *ngundu* (dancing headgears) were brought out.

The *maridzangoma* (expert drummers) and *vanagwenyambira* (expert *mbira* players) set store, *makate* (beer pots, singular *gate*) came out, and the community danced late into the night. As each individual at long last retired home without any goodbyes (for *adhakwa haaoneke*/a person who is full/drunk does not say goodbye), they sang discordantly along the way home, staggering this way and that, their voices dragging, perhaps imitating the equally complicated striding of their beer-wizened feet.

Conclusion

The Kenyan novelist Ngugi wa Thiong'o once said that the work of theorizing from within traditional knowledge involved self-examination in search of a compass with which to navigate outward and discover worlds and peoples around ourselves. Addressing this treatise on *Decolonising the Mind* to African intellectuals, he implored us to place Africa "at the center of things, not existing as an appendix or a satellite of other countries and literatures, [but rather that] things must be seen from the African perspective." In a particularly illuminating paragraph, Ngugi declared:

The oral tradition is rich and many-sided . . . the art did not end yesterday; it is a living tradition . . . familiarity with oral literature could suggest new structures and techniques; and could foster attitudes of mind characterized by the willingness to experiment with new forms. . . . By discovering and proclaiming loyalty to indigenous values, the new literature would on the one hand be set in the stream of history to which it belongs and so be better appreciated; and on the other be better able to embrace and assimilate other thoughts without losing its roots. (wa Thiong'o 1986, 95)

Ngugi was saying that the roots of orature were to be found in the lives of ordinary Africans, and in their songs, compositions, and art (wa Thiong'o 1986, 95). However, like Mudimbe, Ngugi had in mind at the time an "ordinary" African who was a mere informant while orature was just a source of evidence or archive that could inspire the real intellectuals— us!—to be more theoretically original and interesting. His approach echoed Jan Vansina's use of everyday language, oral expression and discourse as an archive for understanding "the dynamics of everyday life, the concrete" (Vansina 1986). Yet as Masolo observed, Vansina and others like Henry Odera-Oruka (1983) went only as far as seeing Africans as storytellers and therefore "a mere resource material from whom the scholar extracts and constructs his mute knowledge." Such an approach enabled the Africanist to cover himself in glory as "the systematic thinker . . . who wades through the ignorance of his interlocutors in order to sift out episteme from *doxa*" (Masolo 1991, 1005).

For Masolo, Mudimbe does not help us in addressing this problem either because he "does not even slightly indicate what he envisages as the relationship between the stylish, scholarly, and elegant deconstructionist method and the idioms of everyday life which embody the mental (epistemological?) schemes of 'traditional' discourse" (Masolo 1991, 1005). Mudimbe insisted that he had no intention to do that, but only wanted to create a room in which others could "dream about another practice for Africa." In undertaking the project of appraising us on the risks of appropriating Western social science, he was searching for a space where Africans could justify themselves "as singular beings engaged in a history that is itself special" (Mudimbe 1973, 10; Jules-Rosette 1991, 947). They could get in there and tell their own stories of innovation while taking stock of their shortcomings, as opposed to being trapped and suffocated by a Western epistemology whose favorite subject was the victimized African who could only find salvation in the West's benevolence. A knowledge acquired "in this conscious journey through the experience imposed by the other" was the key that made it possible for Africans to "operate a critical return [for] without this return which is also a departure . . . it is impossible to think of an after" (Jewsiewicki 1991, 963).

This chapter establishes one foundation upon which an epistemological return can be undertaken; moreover, it is also a departure in broaching guided mobility as a governing motif of such a return. If there is one thing colonialism and slavery did to near perfection, it was to muddy indigenous knowledge with the black brush of village and countryside as autarky and backwardness while depicting the cities and Europe as the designer space for all that was civilization and modernity. The urban focus of Africanist scholarship, both as represented by the research topics of faculty as well as students, and the tendency to prioritize the colonial and postcolonial periods, reinforces this. It is to this village space that I now turn to illustrate how guided mobility was also, at core, a mobile workspace.

2   The Professoriate of the Hunt

In precolonial times, *kuvhima* (hunting) was a transient workplace through sacred forests and an example of mobility as work. In chiShona, *hombarume* was a hunter possessing a *shavé* (spirit) of hunting. Few matched his popularity in society; "meat was the most highly prized food, [and] hunters who could bring it home were very popular and acquired great fame" (Hodza 1979, 357). The hunter was serenaded as *gara ramasango* (wild beast of the forests), *mapfuranhunzi* (one so expert with bow and arrow that he couldn't miss a fly from afar), *vaZvondozvemhuka* (one ever on the trails of the hooves of animals), and *vaChibipitire* (bearer of great loads of meat).

The hunter was believed to transform into a lion of the forest when in the forests, his *shavé* making him a predator, taker of the lives of other animals. Even a lion respected him. Indeed, this venerated predator became unto the hunter *mukombodzwa* or the great beast who was generous, who did not "forbid others to profit from its kill when it has had its fill" (Hodza 1979, 356–358). As we saw in the last chapter, this animal also symbolized, after all *mhondoro*, the great lion-spirit and ancestor of the clan. Hence the word *mhondoro* was also another honorific name for lion.

The dual meaning of *mhondoro* as lion and spirit returns us to vaShona's philosophy of the forest as a sacred space full of malevolent and benevolent spirits, which necessitated that those mobile within and through it seek the guidance of the spiritual. Thus prior to a hunter's departure for the sacred forests, a ceremony of music, dance, beer drinking, and prayers was held to invite *shavé rekuvhima* (hunting spirit) to possess and turn him into a creature of the forest. A hunter did not just burst out of the village like cattle from a kraal, but gathered with other hunters and approached their ancestors to ask for guidance. They told their forebears that *nhomba yatibata pakuru* (craving had seized them at a big spot) (Nyevera 1983, 32). These *vanamurondatsimba* (master trackers of footprints) asked for very specific guidance as they wandered into the forests (*kupenga namasango*). Could the ancestors open the forests so that there might be something to catch? Could they chase away or hold the mouths of the predators? Pacify the prey and make the hunters' weapons unleash their material properties of lethality? And steady their hands so that their arrows might strike their target with efficacy (Gowera 1983, 161)?

MaTshangana did the same. They did not just borrow from vaShona they found living in Gonarezhou, but also brought with them the philosophy of *hlengwe*—that is, gathering to plan the hunt, of strategizing where to go, setting targets, assembling appropriate means, and asking *mikwembu* (*midzimu* or ancestral spirits) for guidance.[1] It was the responsibility of the chief to choose when and where the hunt could take place. He would know which type of animals at what particular time of the year were dropping their babies (which is how most cloven hoofed animals give birth). It was a serious infringement for any hunter to kill such animals or their babies.

Marking Time: Temporal Mobility

Each and every muTshangana boy was schooled in the *lembe* (calendar, literally "year" or season) and its main markers, so that they would know when to hunt and when to do other chores expected of them as men. These moon phases (months) were calculated according to the germination or tasselling of crops, forest animals mating or dropping their fawn, and fruits ripening, and other phenomena.

Mpala (antelope/November) was time for the antelopes to "drop their young, and the bush seem[ed] alive with the perky little fawns and their anxious mothers." Under no circumstance was a hunter to orphan a fawn, bereave its mother, or widow a lover (Bulpin 1954, 129–30).

Nkokoni ("blue wildebeest"/December) was when two important things happened. One was the giving birth of *nkonkoni* (wildebeest). The second was the annual gathering of elephants at specific rendezvous to engage in what maTshangana found to be a bonanza of lovemaking. The animals rarely fed, the bulls obviously in must, the cows in ovulation, both on their erotic high, engaging in "a curious, shuffling love dance," no doubt their own version of foreplay. And *that!* was only the beginning. These adults would then caress each other to the pool, amid "great fun and making a loud commotion of trumpeting and stomach rumbling," having first attended to the formalities of age restriction, adults at their own pool, adolescents with their grandma in another, the matriarch making sure they behaved with the occasional tusk poke on the side. When these grandchildren had drunk their fill, and bathed, the matriarch led them to the different "water holes, game paths, the best feeding places, and such other points as are of interest to an elephant" (Bulpin 1954, 79). Meanwhile, the adults would be busy having coitus, the mature bulls driving away the younger and ambitious competition. Each pair would separate from the herd, and in the shade of the trees where there was privacy, they made love (ibid., 79).

The whole orgy finished just as *Hoho* (the time of "laughter and festivity"/January) was arriving with the gift of the marula fruit, favorite of the elephant, and the perfect wash-down to an ecstatic month. These kings of the forest (the lion begrudging) got royally drunk from the fermenting fruit, judging by the decibel level of the trumpeting in the dead of night. Women of the village loved *Hoho* because it ushered forth marula fruit, the key ingredient for brewing marula wine; the men then descended upon the beer parties to imbibe the liqui-talents of those for whom they had paid *lobola*.

Thus, hunting was to be conducted in the manner of culling and respect, recalling that animals belonged to the ancestors, who even spoke through them. The opportunities to catch different animal species at their most vulnerable and most bountiful was always there. Without proper regulation by the chief, and strict taboos governing consequences, whole elephant herds could easily be exterminated in the act of lovemaking, giving birth, or being born. Centuries of acquired and received knowledge were available on the annual rates of increase, out of which sustainable yields were calculated. Extinction was out of the question not because the people were too few and animals too many, or because opportunities did not exist, but because people hunted only enough to satisfy their meat, skin, and ivory needs. The hunt was a collective workspace not only because of its composition but also its shared proceeds; nobody hunted for their own selfish needs. Nor was the hunter meaningful as an individual; he was a hunter for and with(in) the community.[2] The hunters going into the forests were, therefore, ambassadors and instruments of the entire community; seeing them on the chase, one sees the meanings and uses of the forest in the community.

Upon reaching the edges of the forest, the hunters also knelt down to inform the spirit-owners of the forest that they were about to enter their space. They pleaded that the animals be revealed to them so that the children's undernourishment and craving for soup might be sated. They then put *buté* (the snuff of the ancestral spirits) on a large tree leaf, that the spirits might enjoy with other spirits as they guided the hunters through the forests. Clapping hands in a respectful *bu bu bu!* sound, the hunters then set off on the spoor to commence their hunt (Nyamukondiwa 1983, 163–164). *Nhetembo* (prayers) like these, in which hunters informed *varidzi vesango* of their intentions to enter before they actually did, illustrate the importance of guided mobility at the center of forest practices (Chitsungo 1983, 165). Those in *nyikadzimu* knew the good and the bad places; they were best suited to place the hunter's foot on spots with no thorns, to lead him where the animals were, and to allocate a kill in quantities that satisfied him (Shumba 1983d, 166–167). Nature was something that did not exist outside the spiritual, or the spiritual outside nature. The meeting of an animal of the forest and a person of the village was, in essence, a spiritual encounter.

So here we are, at the edges of the forest, getting ready to start the hunt. We have asked the ancestors to guide us as we set off in this transient workspace. The ancestors are to be with us at our every location lest danger befall us, lest good fortune desert us. We have opened ourselves to allow *mashavé* to take possession of us. We are armed with our bows and arrows and other weapons of the hunt.

But first, what knowledge, or *ruzivo* in chiShona and *vutivi* in chiTshangana, guides our mobilities through this sacred space and how is it produced, by whom, and where? Second, what weapons give guided mobility its "teeth"? After all, we know that animals do not fall and die simply by the act of us looking at or pointing at them, never mind how powerful our *midzimu* and *mashavé* are. Third, what is the place of *shavé* in the hunt, and what are our obligations to it that will affect our hunt? Fourth, now that we have kept all those obligations, what is our reward—what are the products of our laboring in this transient workspace? Fifth, how do we leave the forest—do we just grab our weapons and the meat, and go home? And do we just burst into the village? Those who receive us—do they just take the meat, utter a few "thank yous," and that's it?

The Professoriate of the Hunt: Four Sites of Indigenous Education

VaShona and maTshangana are the two ethnolinguistic groups that live in the southern end of the Mozambique-Zimbabwe border region. Through processes of migration between 1500 and 1820, three different ethnic groups came to be mixed up: vaShona, who moved into the area following the decline of Great Zimbabwe (1500 onward), maTsonga, who gradually moved inland from the east African coast (1750 onward), and maTshangana, who were fleeing from Shaka Zulu (1820s), and who assimilated maTsonga (Bannerman 1981). This book largely concentrates on the Chipinge-Gonarezhou belt along what became the arbitrary border the Portuguese and British demarcated in 1892. The in-migrations and interactions meant that maTshangana and vaShona borrowed knowledge and practices from eachother, making it imperative to defer to both of their hunting traditions throughout the book.

Hunting is *kuvhima* in chiShona and *kuhloteni* in chiTshangana. MaTshangana use the term *maphisa* (single *phisa*) to refer to *expert* or *professional hunters*. *Phisa* is the equivalent of vaShona's *hombarume*. Any person can be a *xihloti* (hunter or *muvhimi* in chiShona), but *phisa* is the ultimate authority on *vutivi-bya-vahloti* (expertise of hunting, or *unyanzvi hwekuvhima* in chiShona). This expertise is acquired in the course of the hunt and transmitted from generation to generation through oral rendition and practice. The hunt is an example of a transient workspace, where knowledge is not only produced and disseminated, but also drilled into a muTshangana youth through practice, and its uptake verified through successful execution of set tasks. The knowledge is concerned with the natural world and is produced, preserved, reproduced, and refined through empirical observation, practices, and experience.

The concept of a transient workspace serves as an important device to account for the transient work hunters do as they move through the forest. For reasons of time I confine myself to four such spaces corresponding to sections 1, 2, 3, 4 below. The first two sites relate to childhood and upbringing and can be seen as transient spaces of education. The other two are concerned with mastery of the techniques of hunting and survival in the forest. Thus recast as a site of mobile instruction or education, the hunt becomes unfolding, enfolding, and productive movement through a sacred forest. Minus the singing and its pacing effects, the collective work ethos of the *nhimbe* is mobilized (in the sense of being rendered mobile), turning the methodical ways of walking through the forest into combing-through-the-bush movement, whose conduct is subject to spiritual regimes of discipline proper to a sacred space. Forest as spiritual realm is no longer only a space of the hunt but also a presence, a force, even a being that standardizes—or forces the standardization of—all practices through the observance of taboos.

1. *Mahumbwe*

Mahumbwe is recreational activity that involves children acting out the chores of fatherhood and motherhood. In *Chirimo* (the time after the harvests), *vasikana vasati vabubudza* (girls who had not yet exhibited first signs

Figure 2.1
Boys driving cattle from the pastures to the kraal were expected to bring something
to the *huvo*.
Source: TKAV 2008

of breasts), and *vakomana vasati vabuda choya* (boys who had not yet developed pubic hair) played at being *baba* (father), *amai* (mother), and *vana*
(children), each engaged in itineraries of manhood, womanhood, and
childhood that society expected to be proper to them. *Baba* and *vanako-
mana* (sons) went out to hunt and returned with meat. It was expected of
boys to hunt or trap birds, mice, and hares for their mothers, otherwise
the family would eat *derere* (okra) or *mufushwa* (boiled and dried vegetables). Food preparation and serving, a portfolio involving stamping and
grinding of grain into meal and cooking, was the responsibility of *amai*
(mother) and *vanasikana* (daughters) (Hodza 1979, 305).

2. Padare

Padare or *dare ravarume* in chiShona or *huvo* in chiTshangana was the
"men's fireplace." It was usually located out of earshot from women to give
acoustic security to manly discussion. Parallel to *huvo* was *ndilo yavamhani*

(women's fire, inside the family hut); the one was out of bounds to the other. *Huvo* was a transient space capable of being moved about. The principal element ever-present was the fire, around which men gathered. Its heat and light created a convivial atmosphere replete with possibilities. It provided a warm atmosphere ripe for conversation, heat for roasting all kinds of foods, and light for people to see each other and what they were roasting. It was here that grandfathers, fathers, sons, and grandsons engaged in conversation light and deep. Here was the ultimate school where boys were trained to be *real men*.

Here grandfathers imparted the clan's virtues and values, while the uncles (mother's brothers) groomed their nephews for the task of responsible husbandry and fatherhood. Both grandpa and uncle were called *vasekuru*, the grandson or nephew *muzukuru*. *Vasekuru* was also responsible for giving his nephews the medicines required for their development through puberty into men, called *mishonga yemisana* (medicines for the back) (Hodza 1979, 18). In spaces like these, *vasekuru*, the fountain of clan wisdoms, imparted to his *vazukuru* (plural, grandsons or nephews) *umhizha* (ironsmithing skills) if he was *mhizha*, *ugamba* (heroism) if he was a warrior, and many other lifelong skills (Majaya 1983b, 331).

It was at the *huvo* that matters of hunting were discussed. The young apprentices were schooled to see and utilize the forest as a reservoir of meats, medicines, and foods. The most important rule was that each "man" was supposed to bring something to *huvo*—logs for the fire, maize cobs, peanuts, sweet potatoes, meats to roast, and stories to tell. These "meats of men" (*nyama yavanuna* in chiTshangana and *nyama yevarume* in chiShona) or meats of conventionally inedible smallish animals (*svinyama* and *nyamanyama* in chiTshangana and chiShona respectively) were composed of civets, genets, monitor lizards, monkeys, mongooses, springhares, and even lion and leopard when found. The meats were taken straight to the *huvo* as men or boys returned from hunting or herding cattle. It was taboo to bring the meats into the kitchen, for them to be seen or eaten by women. To hide the identities of the animals, women were simply told the meat was *svinyama*. The trappings of a prolific future hunter were already emerging as the boys brought something to the fireplace from their day's itineraries.[3]

3. *Kuronda Matsimba*/Tracking
Following the footprints or tracking is called *kuronda matsimba* (chiShona) or *kulanzelela mikondzo* (chiTshangana). A tracker is called *murondatsimba*

(chiShona) or *malanzela* (chiTshangana). *Gwara* (way, or spoor) is the collective trail of footprints or evidence of the direction of passage left behind when something moves through a space. The spoor was the primary transient school for training *murondatsimba*. A muShona or muTshangana boy began learning the science of footprints very early in life. He was trained to observe peculiarities of every animal or human footprint and specific ways of walking of every member of the village (Junod 1927: II, 55). This knowledge was not only for hunting but also for security against attack from witches and enemy warriors.

Tracking was a process of knowledge application, of studying the footprint and psychoanalyzing the animal, what it was thinking as it moved, why it was taking each step in the way it was doing, and taking a particular line of flight. At the end of the day, the hunter had one objective: to catch up with the animal and kill it (Junod 1927: II, 55). The footprints were not always available or obvious, and the tracker might have to start without them. The footprint was a dialectic outcome of contact between the animal and the ground. Sandy plains enabled antelopes to leave a trail of clear footprints in the soil. If the footprint was dark in color, then the dew had fallen into it, meaning it had been made during the night, or hours before dew fell. Should the edges be sharp, then the animal had passed after dewfall while the sand was still wet and firm. If the edges were irregular or crumbled, the prints had been made at noon when the soil was dry and loose.

On hard or clayish soils, the footprint was much harder to read. The important mark to read was the depth of the footprint. If deeply marked, then the earth was wet and soft when stepped upon. That could only mean the animal had passed through early in the morning, right after dewfall. In rocky terrain, the hunter looked out for a stone displaced or turned upside down by an antelope's hooves. If the upper side was wet, the animal had passed before dewfall. If dry, the animal had just passed and was nearby (Junod 1927: II, 55). Reason and instinct folded into one and became skill: "the sight of the footprint immediately [gave] the hunter the information required." He observed for every sign, interpreted it, and drew conclusions on the spot. The hunter could, when tracking two antelope of the same species, determine whether a male or female had made the footprint since the latter was smaller (ibid.). So astute were the tracking skills of *malanzela* that one European traveler writing in 1895 elevated "the noble art of spooring" into "a science" better left to the "black companion" who guided these European itinerants (Millais 1895, 117).

Figure 2.2
Kulanzelela mikondzo (tracking, literally "following footprint") is something still taught to *maTshangana* children to the present. Here, a *muTshangana* boy "reads" the footprints of a family member through the vlei.
Source: TKAV 2012

4. The Forest as a Space of Ecological Education

We have already seen the forest as a sacred space full of spirits malevolent and benevolent. Thus the forest could give (hence *sango rinopa waneta*/the forest rewards the tired), withhold (*nhasi masango matema*/today the forests are dark, when hunters caught or killed nothing), or get angry (*masango atsamwa*/the forests are angry). Later we will see how a hunter's keeping or breaking of *miko* (taboos) caused these three outcomes; we have already seen how fruitful journey was attributed to ancestral guidance while an unfruitful or tragic one represented the opposite. Fruitful mobility was also one that deployed specific knowledge to understand and get the best out of the forest. This knowledge was taught to apprentices in the forest, on site, and through showing and doing.

In this section I isolate for closer examination the knowledge related to plants, starting with the *ximuwu* (baobab tree) and its multiple purposes. The apprentice was shown and instructed how to crush its fruit, enclosed in a thick-crusted pod the size of a papaya, to expose a yellowish powder that could be eaten raw or cooked into porridge.[4] He was taken inside and shown how *ximuwu*'s giant trunk was hollow—and upon sleeping inside it found how it could be used for shelter. When camping for the night, salt was not a problem; saltpans and salt plants were everywhere.[5] The *ilala* or *hanga* palm, when cut at the bottom, oozed pure natural wine. The *nkuhlu* (Natal mahogany) tree's seeds, when put in water for two to three days, and then drained of the water, were squeezed to produce skin oil.[6]

If the hunter fell ill or got injured while in the forest, he could always count on many medicinal plants. Should he wish to regulate his heartbeat (say after a hair-raising encounter with a charging elephant or lion), he could always boil leaves (and drink the water) of the *mbhandhu* or *mbholovisani* (rain or apple leaf tree). This tree was also where maTshangana conducted rainmaking rituals; when the hunter needed ancestral spiritual intervention to turn his fortunes, he came here, placed some *xinesu* (snuff, or *butê*) on the ground and presented his request before the departed ones. It was taboo, therefore, to burn wood from this tree.[7] If he developed a sore tooth, the hunter could always go to the *ndhulwane* (poison apple) shrub, pluck its yellow fruit, pierce it, and squeeze out the fruity innards and place the juice on the offending tooth. He was not, under any circumstances, to swallow the juice because it was poisonous.[8]

If the hunter developed a running stomach while in the forest, he sought the *nkonono* (silver cluster leaf) tree, whose roots, when crushed,

Figure 2.3
Ximuwu.
Source: Author, 2011

could be taken as an emulsion or in a *ximuwu* fruit porridge.[9] If the diarrhea was bloody, the hunter looked no further than the *xanatsi* or mopane tree. He would crush the bark (the inner lining of which makes strong rope for snares), put the pulp in water to dissolve, and then take a swig. If he caught nothing and the summer was in season, he could always be guaranteed a bumper harvest of mopane worms that fed on the *xanatsi* leaves—a delicacy among maTshangana.[10] Besides being a remedy against upset stomachs, the *nsihani* or giant raisin shrub was also an anti-emetic (it stopped the urge to vomit); all the hunter did was chew the bark.[11] In its stead, the bark of the *nthoma* (or jackal berry tree), crushed and boiled in water, could also be used.[12] If the hunter got wounded, he would go to the *ndhenga* or sickle bush creeper, take its fruit and grind it into powder and then apply on the wound.[13] The forest was therefore a pharmacy for the treatment of all kinds of health conditions.

Skill at firemaking was critical to the hunter, who might spend days tracking an animal. Fire (*moto* in chiShona, *ndilo* or *nzilo* in chiTshangana) was indispensable to camping: to scare away lions and other predators, provide warmth in the cold winter nights, for cooking, and smoke-drying meat and preserving it until it was time to return home. Three ingredients were required for making fire: a small softwood log from the *muwuyu/nkuwa* (fig tree) laid horizontally and held firmly in place; a hardwood stick from the *xanatsi* to bore a hole through the softwood, in the course of which fire was produced; and dry and powdered animal dung (*ndove*), judiciously dropped into the smoldering hole to nurture a fire. Firemaking demanded the sort of teamwork learned during cattle herding and *mahumbwe*. One person could make a fire by himself, but it was faster if two people combined their efforts. The work of firemaking involved one person holding the softwood log firm on the ground, into which another drilled with the hardwood stick, causing incremental friction that ushered into a fire.[14]

Weaponry: Giving Guided Movement Teeth

Weaponry gave movement teeth. By weaponry I mean two things: first, weapons treated as a group; second, the science of designing and producing weapons. Without them, movement through and within the forest would not be *kuhloteni/kuvhima*. The animal would be tracked down, found, but not killed, and no man could exhibit *vutivi/ruzivo* worthy of the title *maphisa/hombarume*. Every hunter made his own hunting tools, except ones demanding of specialized manufacture, such as metal spearheads, arrowheads, ax blades, and knives, which were the province of ironsmiths. Every apprentice received training to be his own armorer. The instruments made the hunt a mobile place of work; they enabled hunters to transform the knowledge of the hunt into carcasses, ivory, and meat.

Three types of weaponry are discussed here. The first group was composed of handheld weapons, such as bows, arrows, and poisons. The second category was made up of animals that were deliberately deployed as assistive technologies of hunting (dogs), or animals of the forest, which, by their movements, inadvertently became aid(e)s to the hunter. The third group was composed of remote killing technologies, principally snares, pitfalls, and gin traps. The first two categories involved the weapon or its missile leaving the hunter to strike the quarry, the third involved the quarry coming to the killing device. In all three, the hunter was delegating the killing to things detached from his body, including the animal itself, so that it came to meet the projectile or lethal device and, verily, caused its own death by its own movement.

1. Handheld Weapons

With some weapons, the hunter's hands and muscles assumed the role of a trigger mechanism. This was true of *vurha/uta* (bow) and *matlhari/miseve* (arrows), less so with *mukwana/banga* (knife), and *xihloka/demo* (axe). Altogether these were the four basic arms of a muTshangana/muShona hunter.

I would like to focus on the bow and arrow, observing the interaction between the artifact(s) and the work of making and archery. Specifically, how the hunter's pull of arrow and bowstring backward with one hand while the other held the bow firmly or pulled it forward, generated enough stress that, when the backhand was released, the arrow was propelled forward toward the target of aim. These micromobilities delivered the arrowhead to strike into its target, but may or may not have been enough to kill the animal. For that, the hunter needed to augment the lethal power of the arrow by poisoning it. The arrowhead's purpose was, therefore, not necessarily lethality but, rather, to carry and deliver the poison through piercing the animal's skin and staying there. The poison then dissolved into the body, especially the blood and nervous systems, inducing death through cardiac arrest or nervous breakdown.

The muTshangana's bow came in two sizes ranging between three and five feet depending on the size of the prey. Both were made from the wood of the *nsihani* (giant raisin) tree, whose trunk had the most appropriate tension and elasticity. The *ntambo* (string) was made from the skin of a kudu, hartebeest, or bushbuck.

The arrow was a product of ironsmithing, mostly involving vaShona ironsmiths discussed in chapter 1. For example, vaHota got their name from the *hotá*, the curved handle of the *chiféngu* (hoe) that these people used when cultivating yams in the area that European colonizers later called Headlands. VaHota got their name from neighbors in awe of their *unyanzvi* (skill) at *kunanautsa* (smelting) and *kupfura* (smithing) (Hodza 1979, 144). The arrowhead was made of locally mined and smelted iron, the *liselesele* (shaft or handle) from reeds, and the *ntambo* (talons) that tied it from buckskin. The propellant or tail was made with the feathers of *gondo* (eagle) and *gora* (vulture) not just to balance the arrow in flight, but also to imbue it with these birds' predatory powers.[15]

In chiShona the bow was called *uta*, the arrow *museve* (Odendaal 1941, 23). Like maTshangana, vaShona's bows came in two sizes ranging between three and five feet, namely the *dati* or small bow and the *uta* (large bow) respectively. The size of the prey *determined* the size of the bow. Both were made from the wood of the *mutarara, mutswati, mutesa, chiruwari,* and *mutohwe* trees because of their appropriate tension and elasticity. The bow tapered at either end to promote elasticity and arrow speed, while the thicker middle stabilized the trajectory of the missile.

The string (*musungo/mukosi*) was derived from the hunt. Thus the product of the hunt was also feeding into the raw material for the construction of weapons of the hunt. The tanning involved first cutting up the skin into long strips then tied to a heavy stone. The other end of the strips was tied to the branch of a tree. Next a pole was driven in horizontally between the strips, wound round and round and then let go, pulling the rope tight, loose, then tight again, oil being applied all the time. After it had shed its toughness (*kubva hutong'o*) and become thoroughly moist and soft, the rope was then stretched out to dry. It was then cut into string as appropriate (Snowden 1940, 62; Odendaal 1941, 23). This is exactly how my father taught us to make leather rope from skins of cattle.

By its materiality, the Shona arrow brought together three different domains of nature and expertise: the arrowhead or *museve* (mining and metallurgy), the shaft/hand or *rwikiro* (reeds, basketry), and the talons that tied it (skin, hunting). The arrow shaft varied in length (twelve inches to two feet), the tail being fitted with feathers (*manhenga*) of an eagle or other large bird to balance the arrow in flight. The shaft was eased into the head using a process called *kupisira*, which involved pushing the reed stick into the heated and hollowed arrowhead stem (*runje*). To finish, the binding on the arrowhead and tail was enameled using reddish gum from the roots of an unspecified tree. Together, the bow and its arrows were called *utati*. The quiver for putting arrows was called *homwe yemiseve* or *mukutu*, hence *chinamukutu* (the one with a quiver), the other name for *hombarume*. When filled with arrows, *homwe* became *goba* or *nhava* (sling bag) (Snowden 1940, 62; Odendaal 1941, 24).

Only one ingredient, poison, remained; it was prepared in ways similar to those of maTshangana discussed shortly. Most arrows were usually poisoned with different poisons extracted from snakes, beetles, and plants. Arrows were delegated the task of delivering the poison—syringe-like—into the animal. If the Khoisan and the San were known for their beetle larvae poisons, a muTshangana hunter's signature poison was *vutsulu* (*strophantus kombe*), a vine whose seeds induced death through heart failure (Schapera 1927; Low 2007; Hall and Whitehead 1927). The formula was very simple. The hunter ground the *vutsulu* berries into powder. He took the bark of the silver raisin tree and crushed it into pulp, then mixed the two ingredients in water. The hunter then smeared his arrows with this gooey poison and put them out into the sun to dry. A person with an open wound was not to make or use *vutsulu* lest he poison himself.

Vutsulu only affected the point where the arrowhead entered the flesh; it did not disperse throughout the body. The first thing to do when skinning the carcass was to remove the flesh at the arrow's point of entry and burn it to prevent dogs from eating it. Otherwise the rest of the meat was uncontaminated.[16] This was because *vutsulu* killed quickly and lost its toxicity quickly; it acted on blood circulation only and went straight to arrest the heart.[17] Later, Western physicians would insist that if the wound was poisoned and the animal died from it, the rest of its body would be automatically poisoned. Locals rejected the premise: they had eaten the meat and were still alive (Odendaal 1941, 24).

2. Animals as Weapons

Animals too were weapons, moving on their own, some under human guidance, others guiding human movement. A muTshangana hunter turned both domestic animals (dogs) and forest animals (vultures) into tools of the hunt. He deployed the dog in tracking (through sense of smell) an animal that had passed through without leaving any footprints, and, in packs, to chase and kill animals as large as antelope.

A dog was neither pet nor pest; it was a friend when it killed and an instant enemy when it always lost spoor or trailed and failed to catch an animal. A dog-friend was one that was tenacious and deadly against animals; there was nothing beautiful about a dog. To be tenacious, dogs were "treated" with a medicine called *makotso*, which made them strong and fearless. The moment they saw—better yet smelled the scent of—an animal, their teeth and tongues would feel a surge of appetite for biting flesh.[18]

Later we will see how cunning maTshangana elders no longer able to chase could still study carefully when and where local lions hunted, wait until the lions had killed their prey, then drive them off the carcasses and harvest the meat. This keen study of animal mobilities applied in particular to vultures. That attribute of being the forest's undertaker-in-chief—along with the hyenas, jackals, crows, and maggots—was also weaponizable to the muTshangana hunter, for whom the vulture was an invaluable navigational aid. All he had to do was read their airborne movement carefully; if these rather revolting buzzards kept circling, better yet descended and landed, there was a dead or dying animal there (Junod 1927: II, 55). These aids to hunting were never handheld, the dog being within the hunter's acoustic and visual control, the vulture beyond his control, but

Figure 2.4
Gora, the vulture, undertaker-in-chief of the forest, navigational aid to the hunter.
Source: Author, 2011

weaponizable by its movements betraying its intentions to the hunter. Both were, of course, believed to be subject to *Xikwembu* (God) and the forest's powers of providence.

3. Remote Technologies of the Hunt

By remote technology I refer simply to weapons or devices enabling the hunter to kill or capture animals while distant in space. "Remote" here does not refer to the ability to "control," but the capacity by which hunters were able to accomplish their objectives in and because of their very absence. In this scenario, absence acted as a camouflage, as a weapon even; it tricked the quarry into thinking that the hunter was not there, and feeling safe to move around, yet the hunter had delegated his presence to a weapon, camouflaged it and recused himself. The hunter was absent in physical form, but his knowledge (of the animal and how to kill or catch it), his work (of making and setting the trap), and his objective (to kill or catch the animal) were present in the device set. The animal moved about without suspicion—it literally came to meet its own death. The lethal element did not come to the animal, but the hunter set within the device a trigger to go off when stepped on, when the animal tried to feed, or fell on it. MaTshangana and indeed vaShona used at least three such devices.

Goji (chiTshangana) or *hunza* (chiShona) was the name given to the pit at the bottom of which were spikes made from hardwood, especially of the *tlhonga* tree, that were laced with *vutsulu*. It was a mechanism to turn the ground upon which an animal stood or slept into a weapon against it, and to kill or catch it without being physically present. Dug to a depth where the targeted animal fell in without chance of escape, the *goji* was well camouflaged with grass or leaves.[19] Several pits were dug at intervals, the space in between them fenced with poles and branches, with one end left open to serve as an entrance for the animals that were being stampeded in. Along the fence at open intervals, more pits were constructed to decongest the main one. Labor for digging the pits was communal, and not for men only either (Lloyd 1925, 62). The pit provided a propulsion mechanism that dropped the animal toward the *vutsulu*-poisoned spikes at the bottom, bringing it to its death. The target actually came to the projectile, its velocity boosted by the force of the human-made gravity.

Like *nhimbe*, the making of a *goji* demonstrates how African societies undertook large-scale mechanized projects of mass production that made maximum use of human power. Everyone in the community was involved— women, children, and men. Digging the pit and constructing the fence were labor-intensive tasks. They demanded a lot of energy. As was common practice in southern Africa, women and girls managed all food preparation and distribution (Sheldon 2002).

By preparation I am talking about a society in which women were the human- (as opposed to oil and even much later electric-) powered hammer mills by pounding and then grinding grain. Their grinding apparatus composed of a stone platform and a large handheld round stone. Bigger grains like maize were pounded in a mortar using a wooden pestle first to break them to the requisite fineness to make the task of flouring on the grinding platform easier. The energy that powered the digging of *goji* and the erection of the fence came from the hands of the women and girls; all food preparation was their chore.

While some of the villagers were busy digging the pits at designated exit points, others would be constructing the fence between these traps. In thick-forested terrain this was simply a matter of cutting saplings and branches and stacking them along the prescribed fence line. Ultimately, the fence served the purpose of channeling the animals toward the pits. At night grass torches were lit at designated points along the fence to drive the stricken animals toward the pits. Then, amid much singing, shouting, whistling, and drumbeating, the men drove the animals through the fence, toward the openings, straight into *goji* (Lloyd 1925, 62).

The work of digging the pit involved many tools. The pit was either cornered or round, its size depending on how much meat people desired to harvest. It was dug with hoes and digging sticks to a depth where an elephant or giraffe would not be able to get out. The soil was carried away in calabashes, large pots, and dried animal skins, with women providing the head portage, while men focused on digging. With the pit complete, and as all women and children left, men set about laying thin wooden beams across the pit, covering them with grass and soil to resemble the surroundings. They then inserted the spikes at the pit's bottom, laced them with *vutsulu*, and clambered out to the surface. The final act was the spiritual and medical arming of *goji*, involving the smearing of the whole place with *ndzedzena*, a medicine to make an animal forget and "to circumvent its wariness" Junod 1927: II, 58).

Other remote killing methods were less labor-intensive and directed at killing only one animal. *Nkatsi* (chiTshangana) or *dhibhura* (chiShona) killed by yanking its victim into the air, suffocating it to death. Called the gin trap by *valungu* (*chiTshangana* for "white men"), *nkatsi* was made out of a sturdy sapling bent over and attached to a snare, which was pinned to the ground by means of wooden sticks. Bait was then placed enticingly at the point of strike. When the animal came for dinner, it was yanked by the leg or neck a few feet above the ground. People were not exempt (Partridge 1958, 117).

Chief Mapungwana's maNdau (also vaShona people) used a particularly effective technique to kill elephants: by simply "making holes the size of their feet, with a stake loosely fixed at the bottom, which runs into the wretched animal's foot and remains there, preventing his moving, until they shoot him with poisoned arrows" (Erskine 1874, 122; Mhlanga 1948, 70).

Uguhu: The Spirit and Practice of Elephant Hunting

All these technologies of hunting might be perfect, but as far as maTshangana and vaShona hunters were convinced, so long as *mikwembu/midzimu* (ancestral spirits) and *Xikwembu/Mwari* were not invited to take charge, neither the weapons nor the skills could yield anything. Even those with *mashavé* required the guiding hand and eyesight of the ancestors. They might go hunting in the forest, but see nothing. At last, in dejection, they beseeched their departed yet everpresent forbears: *Ndimi munotipa maropafadzo ose* (Only you can give us all blessings) (Mashiringwane 1983a, 118).

In chapter 1 I pointed out that vaShona use the same word for skill and wandering spirit: *shavé*. *Maphisa* or *hombarume* exhibited such creativity, illustriousness, or industry. In chiShona, when a person has informed the ancestors before embarking on a journey, the elders will say: *Rwendo rwuri kutungamirirwa nevadzimu* (The journey is being guided by the ancestral spirits). However, even fools, cowards, lazybones, average or even amateur hunters might ask for ancestral guidance, but never attain the level of success of *maphisa* and *hombarume*. *Shavé*, as we have seen, does not possess one who does not possess or show promise of possessing certain skills, just as skills alone are not enough to attain success in the forest. Elephant hunting is a perfect example.

Among vaShona, a *hombarume* who specialized in hunting elephant was called *maguhu*. The elephant was respected not only because of its size, but its perceived senses of intuition and wisdom. Basically there were two types of elephant, the tusker (*goronga*) and the tuskless (*muvi*), the latter stronger, speedier, quick-tempered with plenty of courage to overcome the former in any fight. *Muvi* fought with its trunk (*murewo*), led the herd, and sired the calves.

Figure 2.5
Goronga, the tusker.
Source: TKAV

The intuitive powers of the elephant were venerated. Hunters said that unlike other animals the elephant had no wisdom teeth, but possessed four wisdom sticks (*mingano*), each about half the size of a matchstick, located in pairs on either side of the temple under the skin. The calves had a single stick in either temple. These wisdom sticks were very important to vaShona, who believed them to be the source of the animal's great powers of intuition. Hunters coveted these sticks to the extent that upon killing an elephant they had to secure them immediately: the power to forecast the results of future elephant hunts lay in those sticks. They were ground into a powder and boiled with certain herbs and lion fat. *Maguhu* leaving for the hunt took this medicine before going to bed on the eve of the hunt so as to induce dreams of elephant. If the hunter dreamt of himself killing an elephant, he would set off the next day or two with a spring in his step (Greaves 1996, 22–23).

When a hunter came across a tree whose trunk showed signs that an elephant had massaged its neck against it, he looked around for its spoor. He was to "shoot an arrow into the track of the animal and it will not be able to get away." And when he found its dung still fresh, he was to gather it into a bag, "and by this means will be assured of securing the elephant" (Posselt 1935, 131). By that measure he would guarantee himself success. Dry elephant dung was also burnt in small dosage, the smoke inhaled to treat nose bleeding. This was still conventional medicine for nose-bleeding in rural Zimbabwe as we grew up in the 1970s-80s.

One of the greatest powers of intuition maTshangana believed the elephant to have was that of knowing of its impending death or that of family. If a bull in the main herd was to be shot, hunters believed that the herd would know who the victim would be intuitively. When the animals sensed that they were being followed, they were believed to kick up the earth to divine who among them would soon fall victim and isolate him (Parkhurst 1948, 49).

Because the elephant was so wise, it could also tell if the hunter pursuing it was in violation of taboos. A hunter who set out on a hunt with grief in his heart could wound but would never kill an elephant. If he met one with a trunk curled about its head, that was a surefire sign that some tragedy had befallen his family since he left home. Should he see an elephant flinging earth over its back he would know that his wife was bathing, which she was not supposed to do during her husband's absence on a hunt for elephant. The tuskless elephant and the cows with calves would charge and kill those guilty of transgressing sexual taboos unless they immediately confessed their guilt to the elephant. No hunter would allow another to accompany him without first ascertaining whether he had been engaged in transgressive sexual behavior. The hunter had to announce to the elephant the guilty person's presence and name before he fired off his arrow. Women who had difficulty getting the baby out in labor were supposed to do the exact same thing or else they would have stillbirth or die in labor (Parkhurst 1948, 49).

The hunter's wife was not only keeper of elephant hunting taboos but also the source of an elephant hunting medicine called *ndzedzena*. She was expected to bake her placenta the first evening after delivery, hide it the following day, dry it the next evening, before putting it between two pieces of a deliberately broken clay pot. Then she was supposed to hang it from the grass roof of her kitchen hut so that smoke from the fire could mix with it. The husband returning from the hunt then ground it into powder, which he mixed with other medicines for treating his weapons. The belief was that any animal hit with a projectile or caught in a trap laced with *ndzedzena* would never escape. So important was the woman's placenta that if a hunter's wife lost it, she would pay her husband a fine of a hoe. In that case the hunter would seek the placenta from another woman made out of his own seed. That meant only one thing: taking another wife, thereby justifying polygamy (Junod 1927: II, 58).

Miko: Taboos of the Hunt

These medicines could only work subject to a total observance of taboos; in other words, the journey in the forest was undertaken under the strictest of codes of conduct. No hunter was to carry any salt, otherwise bad luck would *rangela* (precede) the whole expedition. It would find no game, never mind how good the tracking, or the dogs might inexplicably lose spoor (Junod 1927: II, 61).

Munyu (salt) was sought only after a kill, just as the hunter made a fire when encamping, also after a kill, lest the forest "thinks" the hunter was taking "its" benevolence for granted. Saltpans and salt plants were liberally distributed in the forests. Without these, the people deferred to plant sources for salt. One of these was *hundi* or *mutyora*—seasoning made by means of "soaking the chaff of finger millet and maize cobs in water after the harvest and straining off the liquid" (Hodza 1979, 331). Alternatively the cobs were burnt, with the ash being strained for liquid salt. But the *real* hunters' salt was at the saltpan or salt-bearing plants in the forest, it being taboo to carry *hundi* home.

Possibly the most important taboo was that the woman left home must abstain from sexual activity while her husband was away, otherwise an elephant or buffalo would gore her husband to death (Parkhurst 1948, 49). The wife had to adhere religiously to this injunction not only to protect the hunter from the retribution of wild beasts, but also "to make him a man of the bush through and through" (Junod 1927: II, 62). The hunter too had to forget about homely comforts, especially his wife and sex, otherwise his medicines would get "heated" (*muti wahisa*) when they should remain cool (*muti watitimela*). He had to become an animal of the bush (*wanhoba*), such that fellow animals would not fear and flee him, nor he them. He could then approach a lion enjoying a sumptuous antelope, "kneel before it, clasp his hands," and ask for meat, or simply advance toward it with the fearlessness of one who is not beast. Like a gracious fellow hunter, the lion was supposed to show genuflection, slowly retire, leaving a fellow predator to fill his stomach (ibid.).

Products of Guided Expertise

At the end of the day, everything we have discussed thus far boiled down to the production of a carcass. Out of it arose many itineraries—starting with knives slicing between the flesh and the skin, separating meat from leather.

For skinning the carcass, the metallurgist—and the hunter himself, often one and the same person—had two kinds of knives. There was the large sheath knife (*bakatwa*), carried on the person, "secured with a piece of cloth, string or reim (rope)." Its twin-edged, sword-like blade (*banga*) was hand-beaten while the sheath (*hara*) was generally carved using a pocket knife. The sheath was in the form of two pieces of leather, each slightly gouged in such a way that when placed together, they left room for the blade in between (Snowden 1940, 67).

Along with skins of lions and leopards, elephant tusks were the chief's property. For every wounded animal falling, he was entitled to *nyanga yepasi* (ground tusk, or the one that touched the ground as the elephant lay on the side). The hunter could have the other. That also applied to one quarter of any big animal (Posselt 1935, 110). Ivory played a key role in vaShona and maTshangana cultures. Among the chiShona-speaking vaDuma, it was an important part of a deceased chief's burial. His body was mummified in a special hut called *bomero* (the place of drying) "with the head resting on a tusk of ivory." Like all other horns, the tusk was also a container of the clan's war medicines, hence *gona rezhou,* or "the medicine horn of an elephant" (ibid., 31).

Some animals were caught alive as a hunter's tribute to the chief, to provide royal entertainment or to add to a kind of zoo. Take for example the pangolin (*hambakubvu* or *haka*), the uniformly gray-brown colored, armor-plated anteater. If the hunter found it, he was to surrender it alive to the chief, who rewarded the bringer with a goat (Rahm 1960). The pangolin was placed in an enclosure where anybody wishing to see it had to pay a kind of "entrance fee." The chief would then sing to it the song "*Hambakubvu, tamba!*" ('Pangolin, play!'), whereupon the captive animal walked around on two legs, "as it would in any case do." It would roll itself up in its scaly skin, often dozing off in that position, amusing the audience. When the animal's entertainment value dwindled, the chief slew it for *usavi.* The meat was reserved only for himself and his senior wife (*vahosi*), its scales being taken to the *n'anga* to make medicines "to ensure bumper crops." They were also used to make a concoction for bathing *nhova* (fontanelle), the small, soft, throbbing spot on an infant's head, to make the skull bone strong (Jackson 1950, 39).

The porcupine (*nungu* or *jenje*) enjoyed similar royal privileges, albeit posthumously. Whenever anybody killed this animal, they were to take it whole to the chief. Only in his presence could its quills be removed and its body degutted; the entire carcass was then surrendered to him. The hunter's reward was a fowl or "something of more or less like value, depending on the generosity of the particular chief." If however the hunter regularly bagged porcupine, there were exemptions to the requirement of bringing it complete with quills and innards (Jackson 1950, 39). The chief ate the meat while the quills were used as unit measures of gold when trading with *maputukezi* (as vaShona and maTshangana called the Portuguese) (Phimister 1974, 449). They were used as *tsono* (needles), and when burnt, the smoke inhaled as medicine for nose bleeding.

Mbada or *ingwe* (leopards) and *shumba* or *mhondoro* (lions) were not regularly killed but on the occasions that they were, the hunter had to "hurry and fetch his ruler so that the animal may be viewed where it has died." The chief then removed the *chiombo* ball—the part of the lion's throat where the hairs from the skins of its countless victims accumulated. Locals believed this part enabled the lion to roar (*kuomba*) hence *chiombo.* Chiefs desperately sought this part—along with the head and skin—to give them lion-power over their subjects; when they passed judgment at court, their sentence was often described as "heavy like the lion's roar" (Jackson 1950, 39). The hunter who presented a lion or leopard before the ruler had much beer brewed in his honor, was given one head of cattle, to kill at the party or take home. He drank as much as his stomach allowed, then gave the rest to the *"women"*, as the men who had yet to kill a lion or leopard were tauntingly called.

In the event that the hunter killed *mhofu* (eland), "in no circumstance [was he to] skin it in the absence of the chief." Should the distance to the chief's residence be prohibitive, and the ruler could not walk to the carcass site, he sent his ambassador. In all cases, the chief got the heart and the surrounding fat, as well as "the shoulder and ribs on the side lying on his *nyika* (territory)." The hunter also surrendered to the ruler the skin on the eland's forehead, which was an important ingredient in the making of the medicines of chieftainship (*miti youshe*). The logic was that just as the eland was a big buck, the chief would become big in the eyes of his people.

Specific parts of the ostrich (*mhou*), hippopotamus (*ngwindi* or *mvuu*), and genet cat (*tsimba* or *nyongo*) were also reserved for the chief. If the hunter killed the hippopotamus, again he was to alert the chief. Its teeth and feet were the ruler's property, although it is not clear what he used them for. In the case of an ostrich, the meat could be taken "but the feathers and any eggs found [were] not the hunter's property—even if the egg ha[d] not yet been laid," but the chief's. The hunter would be rewarded with a goat. The skin of a genet cat, "the beautiful spotted cat more often and erroneously referred to as the civet cat," went to *sadunhu* or *hosi* (headman or deputies to chiefs), who took it as tribute to the highest authority in the land. In all cases, the hunter's reward was a fowl (Jackson 1950, 40). The skins were used mostly for the kilts (*mutsha* or *nhembe*) and headdress of warriors, with the tails being tied to the sporran of the kilt (Buntting 1949, 29).

The crocodile (*ngwena, garwe,* or *gambinga*) had much more complex spiritual meaning and material uses. Hunters generally steered clear of killing it lest they provoke the anger of the witch who owned it. In the event that one had to be killed, however, it was not skinned until the chief's arrival for two reasons. First, its *nduru* (bile) was a deadly poison; if a rival roasted, powdered, and laced one's beer (*kuisirwa mudoro*), especially at a beer party, "death follow[ed] in a few hours, with pain and abdominal distension" (MacVicar 1917, 3). Therefore, the taboo was that if killed, the crocodile was to be thrown back into the pool otherwise the rain would not fall. Second, there was believed to be a stone in its stomach near the gall bladder, which was a revered charm (*ndarama*). If one swallowed it, long life was assured. The chief therefore made sure he was the one to swallow it instead of dying from the *nduru*. When he had lived so long and was about to die, every attempt was made to make him vomit it so that his successor might swallow it and have a long life also. In succession disputes, the swallower of the stone was proclaimed into office (Jackson 1950, 40).

The Return of the (Successful) Hunter

As they exited the forest, the hunters went into prayer to thank and acknowledge the ancestors for their spiritual guidance and the forest for its generosities. *Kuzvirumba* (self-praise) was strongly discouraged. Thanks and praises were due to the ancestors who opened the forests, availing the hunters to the places where herds upon herds of animals were grazing, such that they killed to their fill (Madombi 1983a, 169–170). When thorns had pricked him, the hunter's ancestors had plucked them out without pain, and given him wings when inescapable danger beckoned (ibid., 171). As a token of appreciation the hunter presented another liberal portion of *buté*, and implored the spirits never to get tired, but to keep blessing and fighting for them. Then they left on the journey home (Shumba 1983e, 168–169).

Just as ritual, taboo, and hunting medicines were invoked to turn the hunter into an animal of the forest, animality was a state of the forest. Only humans returned home; the predator remained in the forest. So the returning hunter had to be rehumanized before entering the village. He was to stop at the edges of the forest and shout. The community's leader, who was beyond harm to this creature of the forest and its spiritual powers of transcending the human-animal divide, walked to meet this semi-animal (the he-it) there and take "him-it" before an *inyanga* (healer) for cleansing to remove "it," thereby freeing the "he" to rejoin society (Junod 1927: II, 62). When he brought back meat, he was not to give any to *lababihiki* (people with a bad condition), such as women who had just given birth or who were having their periods. To do so would be to defile his weaponry and chase away good fortune (ibid., 63).

The hunter usually returned toward sunset or at night, for to walk around during broad daylight with a desirable thing was to court the envy of witches. The hunter received the elaborate praises of his wives, mother, *madzitete* (aunts, father's sisters), and his sisters. They would recite the clan praises *vachipururudza* (while ululating) and clapping their hands (Hodza 1979, 156). The clan praises of vayeraGumbo, for example, paid homage to their prolific hunters, great beasts of the forests, givers of lard, thickeners of broth, slayers of the big ones, those creatures that roam outside while the village sleeps, walkers of many lands, hitters of the ground, who rest

Figure 2.6
Inyanga/n'anga, traditional healer.
Source: TKAV

only when they sleep. They were *Mhandamakan'a*, or those who demolish distance with their feet in search of animals (ibid., 266–268). Such poetry was uttered every time a son or father of the house brought something from the bush—meat, fish, fruits—or did wonderful work worthy of praise. His mother or wife would recite the prayer in homage to his ancestors and to the *shavé* to continue to open the paths so that more might continue to come.

Zvirevereve (breathless utterances) or *nhetembo dzemugudza* (poetics between blankets) were registers for thanking each other's ancestors as well as arming the clan's *makona* (medicine horns) with power to be effective. The clan spirits of both husband and wife were believed to be present at the act of conjugal intercourse. They were therefore invoked in praises which spouses exchanged in the moment they were seized with passion. The carnal encounter, therefore, was the meeting of two clans, demanding its own appropriate ritual of acknowledgment and thanks. Ancestral help in the formation and protection of new life was also sought, so that the sexual encounter could bear healthy children, for whom lullabies would then be sung to invoke the protection of both sets of ancestors (Hodza 1979, 18–19).

This love poetry fell within *mutauro wechikuru* (mature-speak), its not-so-allusive references to lustful passion developed and encouraged as boys and girls matured and readied for marriage. Recalling that sex in vaShona culture was not mere carnal act but the most important ritual for strengthening the *makona* as well as reaffirming solidarities between marrying clans, the invocation of the ancestors via the great feats of husband or wife suggests that sex was a site of intimate prayer. Listen to the Mapungwana man, thanking his wife both for the powers of conquest over the forest and the quenching of his own carnal desires that had built up in his own little forest in his prolonged absence on the hunt:

Hekani, Matendera.	Thank you, Ground hornbills.
Hekani, VaNyemba.	Thank you, Lady Nyemba.
Hekani, mubvana waChivazve;	Thank you, daughter of Chivazve;
mukadzi wangu munakunaku,	my very beautiful wife,
anenge nhanga rendodo kuzipa.	sweet like the spotted pumpkin.
Kufa kwako, ndiri mupenyu,	If you die, and leave me alive,
Unotovigwa chitundundu chete;	Only your bust upward will be buried;
Chiuno chinosara neni.	Your loins I shall remain with.
Hekani, Tembo yangu yiyi.	Thank you, my dear Tembo.
Hekani, Mbizi;	Thank you, Zebra;
Mupiyaniswa;	The one who is entangled;
Iwe ndiwe wava amai vangu;	You are now my mother;
Ndiwe uchandiziva padumbu	You who will care for my belly's
napaura;	needs;
M-m, mukadzi wangu;	Oh, my wife;
Chisakaseveswa;	You who will never be made relish;
Dai zvaiseveswa nhai,	For if it could be made relish,
Hapana aitimura neko	Then I would never spare the life of
ndikamusiyirira. (Hodza 1983, 160)	anyone who dared to taste it.

In all fairness, the praise was never one-sided. Nor was it true that husbands did not appreciate the sacrifices of the women, which were a prerequisite for the hunt's success. Poetry was also a way of thanking a partner who has left a paramour satisfied, especially after days, weeks, if not months of awayness. Either in the act or after the orgasm, the man or the woman launched into *zvirevereve zvepabonde*, or the discourses (or intercourse) of the reed mat (Hodza 1979, 8). Indeed, the effusive praise was an incitement in itself for the hunter to tarry forth into the forests. It exemplified one instantiation of the powers of men and women over each other, bound within the spiritual, collectively participating both in the guided mobilities of the hunt and partaking of its success in a most intimate and spiritual way.

Conclusions: What Then Constitutes Technology, Nature, and Mobility in the Presence of the Spiritual?

Historian David William Cohen once emphasized how marriage between different communities built alliances that locked together new segments of kinship groups into a broader network of interdependent relationships. African marriages continue in most societies to involve a groom paying bride wealth (*roora*) in cattle or other symbolic means designed to bind different peoples into *hukama* (relationship). The relationship was contractual and enduring and could "operate over distances and over time," so that migration extended rather than severed it. Cohen illustrated the role of women's migration from their clan to be in marriage, especially the power relations such mobilities configured. The reproduction, new kinship, and respect that oozed through praise poetry enhanced a ruler or patriarch's power as well as the woman's (Cohen 1977, 123–130). Cohen's intervention suggests that "the hunt" cannot be imagined outside this bond between clans, between a husband and a wife (or a father and mother), between the ancestors of the hunter and those of his wife, or the synergy between the spirits and the individual and collective abilities of the living. This role of the spiritual in African innovations has been observed elsewhere on the continent, especially in ironworking (Wright 2002).

Subject to the spiritual, the hunt in the forest allows us to go beyond framings of the hunt as simply a dialogue between the hunter and his weapons, and the prey. We are able to get beyond the colonial writers' view of hunting knowledge as simply one of the virtues of the "noble savage." The English in particular had long seen the individual San, Maasai, or muTshangana hunter this way—as a "free and wild being who draws directly from nature's virtues which raise doubts as to the value of civilization" (Wolmer 2007, 31). In a rather condescending manner, hunting knowledge was cast as "a germ of goodness" illustrating the freedom, simplicity, and closeness to nature now lost to urbanizing Europe (Fairchild 1928). The spiritual was flipped so that instead of being the rock foundation, inspiration, and guide to the exploration of new things, it became the face of retrograde superstition and stasis.

As we saw, the hunt was one more space of apprenticeship through doing, a professoriate that put store in practical skills, with the apprentice's uptake of such knowledge being verified through mastery in practice. It began as fun, the boys acting the part of hunters and adult men taking care of their families and reading the footprints of their kin. When they became old enough to accompany their hunter grandfather, father, uncle, or brothers, these skills were perfected on the spoor. While herding cattle or out hunting they learned how to utilize the forest for food, medicines, and weaponry. They also mastered the codes of conduct essential for a successful hunt. These standards, or a set of agreed-upon rules needed to create textual or other objects, were part and parcel of the guidance that configured and ensured the success of the transient workspace; spiritualization or taboo-ization was a mechanism for "standardizing" practice (cf Bowker and Star 1999, 13–16).

The hunt is one of many professoriates in Africa whose central feature was innovation and entrepreneurship. Merchants trained their sons how to run a shop or market, to hawk commodities from place to place, and to bargain well. Such a spirit of enterprise was not just a function of being geographically in the right place at the right time as some scholars have claimed (Dumett 1983, 666). Geographical location did not make people innovative; it provided material and a platform for people to exercise their talents. Therefore, people create opportunities; opportunities do not come before innovation. The incoming thing does not necessarily make Africans innovative; people are already innovative in other ways beforehand. That doesn't mean the incoming counts for nothing; it simply means the people count for something and do not lose who they are in the face of the incoming.

3   The Coming of the Gun

The professoriate of the hunt was one mechanism through which technological objects from outside were integrated into African daily life.
An example of this is the gun, which arrived in southern Africa around
1500 with the Portuguese interest in Angola, Mozambique, and the
Zimbabwe plateau.

Since the 1960s, the debate on guns has revolved around what kind of
impact they had on Africa. There were those who saw the impact of the
gun as nothing short of revolutionary, with musketeers being credited with
becoming the decisive factor in war (Kea 1971, 213). They produced "a
revolutionary situation," according to one scholar (Smaldone 1972, 596).

Others doubt it. If it is true that guns caused a revolution in Africa
following their arrival circa 1500, what material qualities did they have
that enabled them to do that? Some critics say that the guns were highly
defective to the extent of bursting in the hands of their firers (Kea 1971,
203–204). Others argue that some societies eschewed guns for their own
indigenous hunting and defensive equipment, notably bows, arrows, and
poisons (Echenberg 1971, 249).

In yet other societies, guns were integrated into these indigenous
hunting and fighting repertoires. That meant submission to the gods,
the ancestral spirits, and *mashavé*. Under the spell of the spiritual, the
enemy's guns were believed to be rendered ineffective; the efficacies against
enemies and forest animals were enhanced while the prey of the human and
animal variety were bamboozled before the gun (Fisher and Rowland 1971,
233–234). Being pieces of metal and wood cleverly put together, the gun
could be disassembled and reassembled, studied and understood. Through
interactions in use, the hunter or warrior gained insight into the gun.
Already specialists in mining, smelting, and smithing—forms of spiritually
guided workings of metal—*mhizha* (ironsmiths) extended their *unyanzvi*
(skills) into a new domain: that of *mapfuramagidi* (gunsmiths). They were

not just "tinkermen" appropriating incoming technology and engaging in "bricolage" (Levi-Strauss 1966). Rather, they even bought defective guns that European traders had condemned and restored them. Some even began to make their own guns from scratch, albeit using samples from Europe (Legassick 1966, 111; Kea 1971, 205; Campbell 1987, 409–410).

This variegated picture of the relationship between the indigenous and the incoming begs two questions. First, why and under what conditions do some incoming technologies acquire salience in Africa while others do not? Second, how do we explain this salience from the perspective of Africans as well as the material properties of these technologies, when both are subjected to the authority of the spiritual?

If chapters 1 and 2 were philosophical chapters exploring the guidance of the ancestors on *nzira* (ways and means) to things, this chapter is a historical exploration of such practices with respect to guns. Its main contention is that the behavior of guns in this specific space can only be fully understood within the professoriate of the hunt. The gun is seen not only as technology transferred, or understood merely as the overflowing cup of Western modernity, traveling and technically configuring, but rather as one more ingredient that gets inserted into indigenous trajectories of doing and becoming.

When the gun arrived in maTshangana or vaShona villages, therefore, it did not ride rough shod over indigenous weaponry but was subjected to local protocols and purposes. It was not just "a gun" but "a hunter's gun," and one among many weapons. It could not therefore work as an abstract entity. For a start, the hunter's gun was not just a material artifact or its workings simply material outcomes of technical design. The belief among hunters was that what made artifacts weapons were not people; rather, it was the *mikwembu* (ancestral spirits) and *svikwembu* (spirits of the forest) that, when consulted, armed guns and turned them into weaponry.[1] The subsections that follow discuss the various encounters of the professoriate of the hunt with guns. Figure 3.1 shows the main places and peoples referred to throughout the chapter.

Part I: The Professoriate and the Coming of the Gun

The generations of firearms south of the Zambezi up until 1897 have been well summarized in Anthony Atmore, Mutero Chirenje, and Stan Mudenge's article on south-central Africa in 1971. Referring to the Ndebele and vaShona in particular, the authors stated the following:

The variety of guns was truly impressive. While muzzle-loaders dominated the Shona collection, the Ndebele possessed mainly breech-loading rifles, mostly Martini-Henry rifles. Other rifles found among the Ndebele included Sniders, Enfields, and those manufactured by Reilly, Rigby and Gibbs of Bristol, Manton of London, T. Wilson of Birmingham, and Terry and Calishers. Among the Shona were found Tower muskets, flintlocks, and old blunderbusses. Although many of these guns were second-hand and often dangerous to use, prices paid for guns were high, ranging from the equivalent of bride-wealth for a single gun among the Shona to as much as 200 lb. of ivory for a good rifle. (Atmore, Chirenje, and Mudenge 1971, 553)

It is not a priority here to get into detail about the feedback loops between the designers and the manufacturers in London and Birmingham, on one hand, and African importers and the broad range of uses they put them to, on the other. This is delegated to other work.

Here, it suffices to spell out the several generations of firearms in use side-by-side, from elderly matchlocks to the latest breech-loading rifles. Mostly English from the south and Portuguese from the east coast, usually traders, missionaries, hunters, and military officers—sometimes all of these in one—brought and sold them in exchange for ivory, gold, and other products of African industry. From about 1700, firearms were also coming from Dutch warehouses at the Cape, principally matchlocks, wheelocks, and later flintlocks, which resulted in the rise in firearms use among the Khoisan, Herero, Nama, and later BaTswana (Marks and Atmore 1971, 517; Storey, 2008). The arrival of the British at the Cape from 1820 onwards gradually pushed Dutch/Boer communities northward, culminating in mass-migrations in the 1830s into lands inhabited by the BaTswana, Ndebele, vaVenda, and maHlengwe.

As noted elsewhere (Mavhunga 2003, 207), the ensuing period from 1840 to 1870, the Boer republics resulting from this trekking (migration) and the British South African Colony anchored at the Cape and Natal had a turbulent relationship that made it difficult, if not will nigh impossible, to prevent individual traders from selling firearms to Africans. Most of the rifles were of the Baker, Martini-Henry, and Enfield type, alongside the more numerous, cheaper, and outmoded flintlocks. Thus when I refer to guns in African hands in the 1800s up through the 1880s, I mean mostly flintlocks.

Arms control generally failed because the laws mandating that all arms were to be stamped and registered prior to leaving the ports of Natal, Port Elizabeth, and Delagoa Bay were ineffective. As Sue Miers (1971, 572) found, "no one government controlled all the routes by which arms could enter southern Africa, [and therefore] one power acting alone could not hope to end the traffic and might simply damage its own interests." Hunters, missionaries, traders, and explorers often brought "gifts" of firearms and ammunition to African rulers as a kind of customs duty or fee to pass through their territories (Atmore, Chirenje, & Mudenge 1971, 547-8). The merchants and indeed the local political authority on the ground succumbed to the demand and good profits; if the Portuguese were selling guns, the British enforced the gun control at their own loss of finance and the potential to make African allies. Moreover, even assuming that they exercised a particle of control over these traders, maTshangana and vaShona (including vaVenda) still hunted elephant and exchanged ivory for guns with the Portuguese at Laurenço Marques and Inhambane, and the Transvaal *boers*. They could also travel to Natal sugar plantations (since the 1840s), Kimberly diamond mines (1867 onwards), and Witwatersrand gold fields (1886 onwards) to work as miners and acquire firearms and ammunition. How they did this and to what ends they deployed these armaments is the subject of the rest of this chapter.

Chinguine, Maputukezi, and maTshangana

Before the European partition in 1890–1892, the areas that now lie within and around Gonarezhou National Park had a tradition of "contract hunting" (hunting under contract) that had begun at least in the early eighteenth century. By the 1720s, Dutch merchants at Delagoa Bay were sending maTsonga caravans beyond Phalaborwa, north into Gonarezhou, because of the power of baPedi traders, who insisted on playing an intermediary role between the coast and the hinterland. However, with maTsonga expansion inland from the south from 1750, the volume of trade increased in the Dutch's favor (Parsons 1993). *Maputukezi* were without doubt maHlengwe and vaVenda's most reliable coastal and overseas market for ivory in the late eighteenth and early nineteenth century (Junod 1927: II, 60).

By the 1850s, the Afrikaners' *jagveld* and *maputukezi's terreno de caça* (their terms for "hunting ground") intersected on the Gonarezhou area, which at that time was infested with tsetse fly deadly to cattle but endemic to forest animals (Preller 1922, 12). To vaVenda and maTshangana, Afrikaners were *mabhuru* (*bhuru*, singular), their own pronunciation of the Afrikaner's other name, the *boer* (farmer). By this time maHlengwe had been in Gonarezhou for a century; the forests themselves were part of Hlengweni, or the lands of maHlengwe, the people who gather to share.

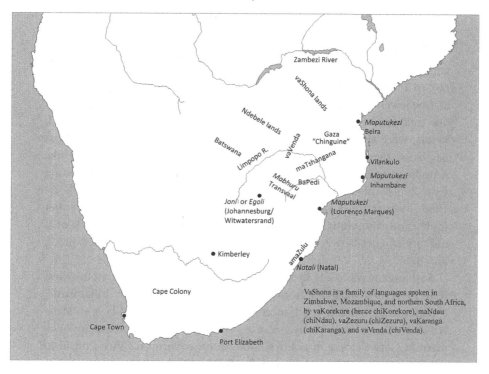

Figure 3.1
Map of nineteenth-century Southern Africa showing the position of the transLimpopo "hunting grounds."
Source: Author

Mabhunu were "the first to extensively apply" guns to hunting in the trans-Limpopo, even though the Portuguese had been around a century longer (Junod 1927: II, 59–60, 140–145; Smith 1969). This is because *mabhunu* were self-governing white settlers while *maputukezi* generally paid tribute to chiShona-speaking kings of vaRozvi, vaManyika, and vaKorekore like every other commoner (Mudenge 1988; Bhila 1982). Two main routes from the east coast, including one from Inhambane, passed through *Chinguine*, as *maputukezi* merchants called Hlengweni (das Neves 1879, 122–123).

These traders acquired immense power in the Transvaal on account of their wealth. In 1859, João Albasini became Superintendent of Natives in the Transvaal republic of the Zoutpansberg after creating a powerful armed force of hunters out of vaTsonga refugees fleeing the Gaza settlement on the lower Limpopo. The Gaza were themselves a Zulu-speaking group fleeing the ambitious and violent territorial expansion of the Zulu king

Shaka. The leader and founder of the Gaza kingdom, from whom maTshangana get their name, was Manukusa Soshangane. But I digress.

With this force Albasini established a staging base for ivory hunting and trading on the Sabie River (in the Transvaal) in 1846 (Liesegang 1967, 66). In 1853 he moved into the Zoutpansberg and set up his fortified base at Goedewensch. Here he settled with his maTsonga retinue and carried out caravans between "Beja" (Vhezha, as vaVenda and their homeland, Venda, were also known) and Delagoa Bay (Wagner 1987, 325). Delagoa Bay-based merchants aside, there were Inhambane-based *maputukezi* like Antonio Augusto de Carvalho and Santa Rita Montanha, both trading with Zoutpansberg and certainly passing through Chinguine in 1855–1856. By that time, Delagoa Bay had supplanted the commerce of that route, and Montanha was coming instead to explore possibilities of a military alliance between *maputukezi* and *mabhuru*—without success (ibid., 325–326).

Soshangane was ready to allow exploitation of Chinguine from the West if it prevented *mabhuru* from allying with *maputukezi* of Inhambane, a policy that his sons and successors Mawewe (1858–1861) and Mzila (1862–1883) changed partly because of Zoutpansberg elements' involvement in Gaza succession politics. Albasini (now Portugal's Vice-Consul to the Transvaal) sheltered Mzila after his flight from Mawewe, as *mabhuru* of Zoutpansberg and *maputukezi* of Lourenço Marques drew up plans to partition the Gaza kingdom (Liesegang 1975). Chinguine was caught up in these interlocked, conflicting interests: in 1860, Albasini marched with his heavily firearmed maTsonga force into Chikwarakwara and exacted tribute in ivory from the chief.

Meanwhile, Mawewe was determined to enforce among his Hlengwe subjects his age-old right to the tusk that hit the ground first as the elephant fell—that is, one tusk for every bull shot. It was the Delagoa Bay-based *muputukezi* merchant Fernandes das Neves's trip to Chikwarakwara (and his rather hasty departure and narrow escape) to the safety of the Transvaal that inflamed hostilities. That very same December 1860, Mawewe sealed the Limpopo boundary and banned all white traffic from the Zoutpansberg (das Neves 1879, 79–82).

When das Neves left Albasini's house on November 2, 1860, he passed through Venda country to see Mzila, Mawewe's brother and rival to the Gaza throne, then a refugee among vaVenda neighbors of Chikwarakwara, who paid taxes to *mabhuru* of the Transvaal. Das Neves was thinking of aiding Mzila in a war to put him on the Gaza throne, thereby guaranteeing access to Chinguine and unlimited ivory supplies to *maputukezi* of

Lourenço Marques. Mzila recognized this as an opportunity to secure guns and powerful allies with which to seize power. Mawewe's blockade of trade routes and constant attacks on Lourenco Marques, coupled with his imminent march on *maputukezi*'s ally Magude made this a worthwhile investment to das Neves (1879, 117–20). The plan went horribly wrong and das Neves escaped through the eye of a needle, but a few months later Mzila did in fact depose his brother and take the throne.

Mzila had intended to use *maputukezi* to get to power and, once on the throne, to assert Gaza authority over the hunting grounds. That is exactly what he did. In 1862, he turned the tables on Albasini and his *maputukezi* merchant allies in the Transvaal and at Lourenço Marques (das Neves 1879, 79–82). Albasini was prepared to accept Mzila's terms if it could allow him to continue hunting, but attempts at rapprochement failed. From 1865 to 1867, *maputukezi* and *mabhuru* were almost continuously in *laager* (all-round defense), not only because of Mzila's military activities, but also because of vaVenda (Wagner 1987, 328; Bonner 1983, 99).

Swart Skuts: Black Marksmen and *Mabhuru*

Mabhuru power rested on a monopoly of guns and their proliferation. By 1860 many vaVenda had acquired guns from these newcomers, who came in initially to trade, before increasingly getting too powerful and asserting themselves as colonizers (Stayt 1931, 74–76).

Even after *mabhuru* became more aggressive and annexed much of the Zoutpansberg, vaVenda and maTshangana remained strategically positioned along the Limpopo. Few white men could risk the mosquitoes and tsetse flies to hunt elephant for ivory in the valley, and vaVenda were reputed elephant hunters. They killed the animal by stalking it, then darting in from cover to hamstring it with axes and immobilize it, as the rest of the hunting party closed in with spears to finish it off. All tusks were submitted to the chief of the particular area, who exchanged them for guns, cloth, beads, and other goods *maputukezi* and maTsonga's caravans brought from Delagoa Bay. VaVenda were also master shots with rifles, and became particularly good at using them to kill leopard remotely. They did this by extending the gun trap technique to flintlocks and later breechloading rifles, tying the trigger with a string stretched across the animal's known path, so that it pulled the trigger on itself (Stayt 1931, 77).

Mabhuru came to depend on vaVenda and maTsonga professional hunters whom the chiefs tributary to the Transvaal sent to render services as a way of paying taxes. Many such *swart skuts* (black shots, as *mabhuru* called them) were young boys captured during slave raids that *mabhuru* disguised as "apprentices" to escape the scrutiny of European antislavery laws governing areas in Africa under white occupation. These boys, besides performing other farm chores, were trained and sent as *swart skuts* to hunt elephants and bring back ivory from the Lebombo Mountains near Punda Maria and north in Chinguine in return for their freedom. In his memoirs, Hendrik Struben elaborated on this Afrikaner system of handing out guns to African hunters:

In the early days of Schoemansdal (a trading village under Zoutpansberg) there was a recognized system under Government permits of supplying native hunters with guns, ammunition, blue salempore, brass wire, and Venetian glass beads. Each hunter according to his recognized value was given a certain number of carriers to take his truck in and the ivory out, and the hunters got a percentage on the ivory delivered. Some of these men were good elephant shots and made lots of money. Many Boer hunters went themselves, taking bearers with them. It was a hard life, fraught with danger. (Struben 1920, 86)

Mabhuru took Africans with them to tap into the indigenous ecological understandings of the local environment, the high (foot) mobility of vaVenda and maHlengwe in tsetse-infested areas neither ox nor horse could reach, and their skills at shooting (hence *swart skut*). Between 1840 and 1867 *mabhuru* distributed guns to groups of these hunters, to be handed over after every hunt, under the agreement that their chiefs would see to their collection. One English trader describes the mode of *swart skut* elephant hunting as follows: "The Hunters all go out on foot with six, eight, or ten [Africans] along with them carrying a kettle, small bag of coffee and biscuits, and a couple of elephant guns from four to eight lb. Their hunts last from ten to thirty days at a time. It is certainly hard work for them, especially if they are unfortunate and bring back nothing."[2] This is how *mabhuru* and their *maputukezi* merchant allies built Schoemansdal into the foremost inland market south of the Limpopo, drawing traders from overseas as well as the African interior.

The *swart skut* was a product of *mabhuru* called the *maatskappy* (hunting guild) that developed into a mercantile capitalist class that owned and loaned hunting equipment to Africans and took the ivory and *biltong* (dried meat, as *mabhuru* called it) harvest as repayment. Entrusting guns in the hands of Africans twenty miles on the other side of the Limpopo was always going to be a big risk. In 1851, for example, Albasini kitted his "apprentices" with guns to provide a protection force for his carriers, up to five per caravan. The Transvaal law governing hunting passed in 1858 stipulated that a white man must accompany any African going into the forests with a gun; that the gun and its African user must be registered; that African hunters who "strayed" from their white supervisor in the bush must be found before night fell; and that for every one white man there must be no more than two *swart skuts* accompanying him. But the law was willfully ignored (Wagner 1987, 331–332).

The notion of "contract" or "apprentice" labor derived from the Transvaal labor laws. There was tribute labor (service rendered in place of a tax) and then there was *apprenticeship* (captives turned into trainees) (Delius and Trapido 1983; Harries 1981). On paper *inboekselings* (apprentices) were orphans that the Afrikaner *kommando* [volunteer milita] would have "rescued" from sites of combat; in practice, it was simply a cover for captives turned into servants. Besides captives, therefore, *swart skuts* were usually tribute labor that vaVenda chiefs seconded to the Afrikaners. After the labor equivalent to the tribute had been rendered, the *swart skut* returned home armed with skills learned among the Afrikaners, such as mending gunlocks, making bullets, and so on. Chiefs and headmen were responsible for distributing to and collecting back guns from the *swart skuts* (Wagner 1987, 333–334).

VaVenda became the conveyor belts for guns and gun knowledge (technology transfer) from *mabhuru* to their vaShona and maTshangana neighbors in the north and east. The best example is that of Ishe (chief) Madzivire and vaMhari people, for whom the adoption of guns into local military architectures paid huge dividends. In the 1860s Chivi, another chief of vaMhari people, paid tribute to the Ndebele. From 1870, under Chivi's successor Mazorodze, vaMhari engaged in brisk trade in guns with vaVenda in exchange for cattle. Guns increasingly became instruments with which to protect these herds from raiders (Wallis 1945, 152; Wallis 1954, 69, 190).

VaMhari not only became skilled users of guns but also *nyanzvi* in *kupfura magidi* (forging and repairing guns) and the making of bullets for them. They had mastered the techniques of forging a variety of guns, two of which they called: *kororo* and *hlabakude* (Mazarire 2003). Just what these names designate is not (yet) clear, they were without doubt copies of imported firearms known among vaKaranga [another chiShona-speaking people in Southern and central Zimbabwe] as *zvigidi* (Ellert 1984, 57).

Here is a case of people as infrastructure through which technological artifacts, ideas, and practices travel from one culture to another. An instance where mobility is no longer just the mute transport or the smooth flow of things, but an atmosphere of frictions within which tapestries of connections are woven through the moving feet, one pace at a time. The story, the technology, is not (in) the gun; the story is (in) the people who seek out, proliferate, design purposes for, alter, fire, and reap from the gun. Otherwise, the gun is an artifact of human ingenuities, by its own materialities contributing to conditions under which certain modes of action become permissible.

Part II: Technology and Remoteness

With the British South Africa Company (BSA Company) occupation of *Mashonaland* (vaShona's lands) in 1890 (and Ndebele's lands in 1893), vaVenda communities were split between two countries (South Africa and Rhodesia) along the Limpopo, while their Hlengwe neighbors to the east were split into ethnic segments of three countries (South Africa, Rhodesia, and Mozambique). Yet, from a hunting point of view, very little changed: Hlengweni remained a marginal locality far from the major seats of power.

The resulting borderlands acquired sophisticated forms of cosmopolitanism because of the very fact of their remoteness. By remote I do not mean Wi-Fi remote (the "new remote"), but the "old" remote, where the lack of transport (road, rail, or telephone) could mean being cut off from distant places. The suggestion here is that this old remote had and still has an enduring allure by the very fact of its "cut-offness." That is precisely why people who are likely to use guns, people who move illicitly (poachers, smugglers, insurgents), and forest animals are associated with the borderlands. With the advent of colonial partition, all of these elements together constituted a new kind of mixture and interaction, whose constituent elements were drawn from different parts of the subregion and the world.

The Birth of a Remote Colonial Margin
The state's power did not reach everywhere. Most of the southeastern lowveld was not demarcated by Europeans before 1897, and consequently remained state land. Effective African administration in Southern Rhodesia, through the Department of Native Affairs, began only in 1894 (Chauke 1985, 3). Peter Forrestall (whom locals called *Ndambakuwa*, "the one who refuses to budge on anything") was the first Native Commissioner of Chibi District, the vast area southwest, west, and northwest of Gonarezhou under discussion here. He was based at Chibi Station a hundred miles away, too far from the southeastern border for effective control. As a result, maTshangana people of Mahenye, Ngwenyenye, Masivamele, and Sengwe

by and large remained unaffected by the colonial partition. The map (figure 6.3) will be useful for orienting the reader herein.

Instead, Ndambakuwa relied on a few white settlers living in the border area for day-to-day administration.[3] They had powers of arrest and were the eyes and ears of the government, a privilege—or burden—that they often misused.[4] Ndambakuwa also relied on "native messengers" recruited from the local villages to maintain law and order—among their beloved relatives! Other than messengers, African chiefs, who became salaried government bureaucrats, were supposed to assist him, but only a few were submissive or cooperative. Even when they did, poor communications and physical inaccessibility meant that frontier chiefs like Sengwe and headmen (village heads) like Masivamele and Ngwenyenye "to a large degree still ruled as if nothing had happened" and exercised considerable autonomy in spite of their tax obligations to government (Chauke 1983, 6, 9).

The Ndambakuwa era illustrates the sharp contrast between disarmament and gun control at the center and the continued firearms proliferation on the unpatrolled borderlands. Unlike the Ndebele and vaShona, maTshangana had not risen against the administration in 1896–1897 precisely because the most oppressive aspects of BSA Company rule had never been felt on the margins.[5] Ndambakuwa always struggled to raise a sizeable patrol of ten men, barely enough for efficient ground coverage of this sprawling district. The lore of the no man's land dominates writing about twentieth-century Gonarezhou and bears no repeating (Bulpin 1954; Wright 1972, 1976; Thomson 2001; Wolmer 2007). A market existed for all sorts of illicitly obtained goods; to call such goods or activities "illicit" at all is a misnomer because such terms are only relevant in spaces where statutory law and "the State" exist. On the margins it did not.

African hunting survived on the margins of the colonies of Southern Rhodesia, Mozambique, and South Africa because of the continued access—albeit illegal access—to guns and *mhuka* that local people continued to enjoy. The Administration made feeble attempts to increase accessibility and control. In 1919 the Native Department appointed the first Assistant Native Commissioner (ANC) and magistrate in charge of the proposed station at Nuanetsi (a colonial corruption of the word "Mwenezi").[6] Only in 1921 was he finally stationed there. The new but temporary police post at Nuanetsi operated only in winter, beating an early retreat before summer floods and malaria closed in.

In 1924, the BSA Company Ranch at Nuanetsi offered the Native Department land and personnel to build new offices along the Mwenezi River to replace its ramshackle suboffice, only built after four years of fighting red tape (Mavhunga and Spierenburg 2009).[7] It was around this time that the Department of Commerce started considering the establishment of a game reserve in the area. The BSA Company still held much political influence through its extensive investments despite handing over administration of Rhodesia to settlers. Therefore it agreed to this increased administrative presence to secure its cattle investments, Nuanetsi Ranch close to Gonarezhou and, further west, Liebig's Ranch (ibid. 2009).

In 1925, the state established a commission under Morris Carter to formalize the veterinary-oriented divisions of land taking place in the area. The Native Reserves were designed in such a way that they formed a buffer between wild animal (vermin)-infested "Crown Lands" and the European Land. Matibi I Reserve had 286,000 acres and Matibi II 478,000; both now fell under Chibi District. In particular, the area between the Lundi, the Portuguese border, the Limpopo and Bubi Rivers, Nuanetsi Ranch, and Matibi II, totaling 1,8 million acres, became "Crown (State) Land" or "unalienated land" (Wright 1972, 325). Any land that was in the hands of government and that was unalienated could always be open to future redesignation as the state saw fit.

Crooks Corner: A Republic of Absence

Few other places disrupt everything that we know about conventional definitions and interpretations of incoming technology more than borderlands. This absence of the state—with all its signifiers and enforcers—stems from the distance of this place from state capitals like Lourenço Marques, Salisbury, and Pretoria. The face of the state—police, army, district administrators, clearly marked boundaries, and the rule of (government's) law—is visible by its near-complete absence. When the state goes absent, along with its colonializing structures, something always takes its place. A state—or states—of things, a kind, or different kinds, of regimes always take(s) over (Bille, Hastrup, and Sørensen 2010).

The way to understand guns in this specific environment is to ask: What then happens when sovereignty is defined by absences (of the state, urbanization, communication and transport bar natural means), transgressions (forest animals, livestock, and people crossing borders governments have installed to keep them apart), everyday life challenges (the imperative to live and make a living), and opportunities and means of utilizing them (hunting animals for meat and ivory, cattle rustling, informal trade, home-based industry, farming, cattle ranching, ecotourism, and migration in search of employment in the cities)? Since when did these absences become so manifest?

Borders are never always fixed. They are (trans)portable—and transient— "places of socialization, intimacy, evasion, reflection, encounter, negotiation, and ultimately transformation" (Jirón 2010, 69). It is not always the case that people cross the border. Sometimes the border crosses them (first). Sometimes it exists only to those that made it, but is completely invisible to those not privy to the transaction. From outside—from the capitals and cities—the border is seen as a margin. From within itself, it is the center, scene of many activities. Here, the movements that animate everyday life exhibit inattentiveness to the forces that govern the conventional center: the city or seat of power. The border is its own seat of power, even its own state, indeed, *a republic of absence*.

Especially for local inhabitants summarily divided into different colonies with the "borderline" cutting straight through their villages, the borderlands offer a continued unfettered existence that run foul of our neat periodizations of African history into precolonial, colonial, and postcolonial. Here, between colonies belonging to different European states, in these interstices and crevices of Empire, are to be found multiple mobilities—of people, of animals, of birds, of insects, of microbes, of technologies, of ideas, all beyond the gaze of colonial states. For this we look no further than Crooks Corner.

In 1892, the three territories of Portuguese East Africa (PEA) or Mozambique, Union of South Africa, and Southern Rhodesia set as their boundary an island where these three colonies met. By 1910 this little island had acquired notoriety in government circles as a place where those running away from state law and seeking illegal fortune from ivory "poaching" (unlicensed hunting) and illicit labor recruiting for the Transvaal gold mines established new bases. They came from all over the Caucasian world—Australia, New Zealand, Argentina, Brazil, the United States of America, Canada, and Europe, including Scandinavia. In the 1920s these white men engaged in hunting, trading, and recruiting without licenses and had the latest guns and access to ammunition galore.

The island became a haven for such "crooks" because colonial administrators struggled to raise sizeable patrols to pursue them. The police could only patrol in winter when rivers were not in flood and tsetse fly and mosquitoes were not hyperactive. Until 1918, the colonial authorities left the Crooks Corner area to its own devices. The "crooks" made it their "corner" for the convenience of escaping to one country if pursued in another (Bulpin 1954, 24, 108). At the same time, within this no man's land the surviving Afrikaner hunting generation of the late-nineteenth century *mabhuru* and a new breed of hunters provided a small but steady market for those who ran guns.

Flanking Crooks Corner to the east was Mozambique, most of it thick-forested and unpoliced. To the south was the Transvaal, which together with Mozambique, was the main source of rifles used in the poaching. The Transvaal-Mozambique frontier southeast of Pafuri was easily crossable on foot by itinerants—African men going to and from *migodhi* (mines, singular *mugodhi*), and white men going into Gonarezhou and the Mozambican areas of Machaze, Banhine and Chinzine, to hunt. There was virtually no difference between an ivory trader/hunter and a recruiter. As the law tightened up on the South African side, the buyers of ivory (who were also suppliers of guns) concentrated on the Portuguese side. Most of them were Portuguese *Chefe de Postos* (chief of the post, or district administrators) and storeowners. Officials often took bribes from and gave unlicensed hunters and labor recruiters a pass. In particular, Portuguese policemen, seldom paid and generally independent of central government control, found it easier to accept a bribe than assume the burden of escorting a prisoner through exacting terrain (Bulpin 1954, 47).

The absence of bilateral border agreements between the Portuguese, the British, and the Afrikaners—the former sharing no borders in Europe but neighbors in southern Africa—compounded law enforcement. Rhodesian authorities often complained that the Portuguese police were uncooperative about requests to pursue criminals fleeing justice. It was common practice for Africans and whites alike to commit crimes in one country and use the border as a shield against arrest. Flight or mobility became a weapon of escape, the border a fortress against the statutory weapons of the state. If one was wanted in Rhodesia for illegal possession and use of guns, neither the Portuguese nor South African authorities could repatriate him for trial. Even though joint operations were carried out in 1918, those arrested in one territory could not be brought to justice in the other.[8]

These itinerant border characters had a fluid citizenship not bound by any national laws or borders. Most had been big game hunters before colonization in 1890; they had roamed freely in the interior, and now regarded colonial game laws as unwarranted restrictions. This illicit image is in stark contrast to the usually positive portrayal of late-nineteenth century ivory hunters as champions of imperialism opening up the interior to European technology. People like British hunter Frederick Selous were "the noble big game hunters" who guided the Pioneer Column, the BSA Company's colonizing force, into the lands he had hunted in for twenty years. They then settled down to a life of mining and farming. There were others who found the daily routine on mines, farms, and burgeoning towns boring, the laws too suffocating, and the vast borderland forests liberating. The famously notorious ivory poacher Cecil *Bvekenya* Barnard, his associates Jack Ford (ex-BSA Police, and Charles Diegel (a German drifter whom the local maTshangana called Chari) fit this description: all had come in search of ivory wealth. They hugged the borders and steered clear of police patrols in all three territories keen to interview them for assault, robbery, murder, illegal recruiting, and poaching (Bulpin 1954, 211–217).

Mobilities, Affordances, Cosmopolitanisms

As production on the move, the hunt exemplifies the cosmopolitanization of the forest: geographic remoteness from law enforcement permitted the flourishing of multiple mobilities that might otherwise be illegal close to a police station or in a city street. We can now see why, in this particular case, the incoming (Europeans) could not take over or have the power to configure a new order, and why their guns became one more ingredient in the fashioning of a cosmopolitanism vernacular to the state's absence and the state of absence. This is one instance where the much-gasconaded powers of colonial regimes are not in evidence, a place where multiple mobilities—of mosquitoes, big animals, people, artifacts, as well as temperatures moving up and down, seasons coming and going, along with them floods and dry, sandy rivers—created conditions where multiple individual temporalities and local rationalities inserted themselves into the (un)making of human-animal-place relations, forging a new republic absent of the colonial state. Here, the professoriate of the hunt is seen being galvanized and remodeled into an infrastructure for addressing the burdens of colonial marginality while subverting that marginality into a space where transient workspaces could be set up without the state's molestations.

The mobilities of hunt-related technologies from licit to illicit to licit again is perhaps the most interesting feature of this cosmopolitanism vernacular to the borderland. Outgoing movements of people in search of work in the *migodhi* of Kimberley and Witwatersrand, on one hand, and the incoming movements of firearms, on the other, constitute two interesting intersections of paths. When the state patrol at last pitched up, an interesting game of hide-and-seek took place between illicit mobilities and the fixity of the law. The cosmopolitanism was also racial. Here was a place governed not by the white settler state, but by the biogeophysical environment.

We often associate cosmopolitanism with the city, trawling the globe for high-paying jobs, or cyber-connectedness. Yet here was Crooks Corner, a remote place drawing people from all over the world to hunt. A place of nature-gone-rogue-on-humanity, full of insects deadly to people (mosquitoes and their plasmodia) and, further north toward Rio Save (the Mozambican side of the Save River), transport animals (tsetse flies and their trypanosomes); a place dry as a bone and hot as a furnace in winter; a place without connection to "modernity," the backwater of Empire.

White frontiersmen like Bvekenya could buy guns in Johannesburg gun stores and get licenses without a hustle; they just needed to produce the money and their white skin. Everywhere, a white man was seldom asked questions when he moved about bearing arms. The fact of his whiteness was the law in itself, confirmation being academic. By contrast, except when wearing a "native police" uniform or in the company of a white man, an African was promptly arrested for illegal possession of arms. On the "invisible" margins, the state had to show up first. The white hunter-merchants bought and issued guns to their African employees, who kept them as long as they hunted together. That gave Africans access to the latest and most effective weapons, especially since the white *master* employing them received no benefit if elephants escaped.

It is important to point out that even with the demise of Bvekenya and his Crooks Corner generation, the forests remained and ought to be seen as a cosmopolitan space teeming with hunters from different nationalities and bringing different types and calibers of guns. This was enhanced in 1937 when, after years of banning hunting on the eastern side of Gonarezhou between the Rio Save and Limpopo, the Portuguese government opened the area to unlimited elephant hunting by safari tourists. Previously, elephants had become a scourge of villagers in that stretch of country from the Limpopo through Jivinjovo and Potsakufa to the Save. These denizens, moving about *sans frontières*, became pests in maTshangana villagers' lives and fields, marauding through their crops, leaving them starving. So in 1937, the authorities threw open the whole forest east of Gonarezhou to safari hunting. There were only two areas prohibited: east of Pafuri near Kruger National Park's northern end, and the Maputo Game Reserve to the far south. Anywhere else all game was fair game (Capstick 1988, 71, 88).

The ban now lifted, hunters of all shades of white descended on the forests. Where gun possession and hunting had been illegal prior to 1937, now there was an avalanche. In fact, the Portuguese administration led by example. The *Chefe de Postos* started issuing Martini-Henry rifles to African hunters based in the villages to contain the elephant problem. However, because of the free-for-all and the ivory rush, care was not taken to ensure that wounded animals were tracked and killed. The country now teemed with wounded lions, elephants, and buffaloes, ready to charge at every opportunity, and trampling their victims into pulp. The agreement was that the *Chefe de Postos* would take delivery of all the ivory and Africans the meat, which was highly lucrative business. Not to be outdone, the mostly Indian storekeepers at Masenjeni, Machaze, Chikwarakwara, and Mapai supplied the hunters with very good rifles and offered to buy their ivory harvest (Capstick 1988, 89).

But maTshangana living in the area were not just "recipients" of guns; they also extended their age-old contract labor practices to help themselves to the opportunities that the white hunters coming into their territory availed to own guns. A gun that was one day being used to shoot sharks in one part of the world found itself facing down the neck of an elephant in Gonarezhou. Africans—particularly maTshangana—served as translators, carriers, guides, and trackers for these incoming hunters since many had relatives in the areas of Chikwarakwara, Machaze, and Masenjeni. Some of these maTshangana came from Lourenço Marques, Bilene, and Vilankulos.

At Chinzine, the hunter coming from South Africa was required to check in at the *Chefe de Posto*'s office before proceeding into the hunting grounds of Machaze to the north. It was necessary to let the *Chefe* know because "sooner or later, his men would advise him that it was a gunrunner or a poacher or some other form of dubious character. So, you'd drop by for *chá* and advise him you were now in his area" (Capstick 1988, 74). Of course, such outward shows of the rule of law were necessary to mask deals that could be struck between gun smugglers and the *Chefe de Posto*, with the visit acting more as a delivery run, and the offices of the administrator as a warehouse for gun trafficking (Bulpin 1954, 212).

In the hinterland of Rhodesia, the Lee Metford, Martini-Henry, and Mannlicher rifles could only be accessible to Africans hunting with and for a white man, while all Africans were disarmed completely. On the lowveld margins, things were not so black and white precisely because of the symbiotic relationship between the guns the white man brought and the understandings and technologies local maTshangana possessed, without which the guns might be entirely worthless as hunting instruments.

American and British Automobiles in Remote Places

Beginning in the late-1920s and early 1930s, more and more of these white hunters came to the lowveld driving trucks and cars. Most of them initially brought their automobiles with them on board ships, disembarked with them at Cape Town, Port Elizabeth, Natal, or Beira, then set off inland to enjoy the scenery of the Victoria Falls (whose real chiTonga name is *Mosi oa Tunya*, the smoke that thunders)"discovered by Dr. David Livingstone," see the Matopos, "where Cecil John Rhodes, founder of Rhodesia, was buried," and the "ruins" of Great Zimbabwe, "built by the Phoenicians." Unlike trains confined to fixed railroads, cars allowed for flexibility to go "off the beaten track," pass through the countryside and experience "how the primitive natives lived," and go off to the outer reaches of the colonies to "shoot wild life with gun and camera" ("Motor Touring" 1931; "Shooting Game" 1934; Tomlinson 1936; "African Grand Tour" 1937).

The "Big Three" American automakers—General Motors, Ford, and Chrysler—dominated the southern African car market. General Motors (GM) South Africa Limited was founded in 1913 and initially distributed Chevrolet automobiles before switching to the manufactured and distribution of all GM models from 1926 onward. Ford came later in the 1920s and set up a factory at Port Elizabeth that assembled car kits from Canada and, later, from Great Britain and Australia. Trucks such as Model T (from the 1920s onward) and then the Model B (beginning in the 1930s) were among the locally assembled automobiles. British cars included brands from Austin Motor Company Ltd. of London, Willys-Overland of Stockport, Vauxhall Motors of Luton, and Morris Motors Ltd. of Oxford ("The Civilizing Influence of Roads" 1929–1930, 144). The government statistics for 1934 illustrate clearly the dominance of U.S. automobiles in southern Africa. Out of the 1,722 private motorcars registered in Southern Rhodesia, a total of 1,407 were American while just 308 were British. In South Africa, 2,092 of the 2,366 automobiles were American while only 254 were British. The figures for South Africa were 627 Chevrolet, 545 Ford, 343 Plymouth, 185 Terraplane, 156 Dodge, 88 Studebaker, 86 Chrysler, 68 Willys-Overland, and 62 Buick (all American), and 104 Austin, 59 Vauxhall, and 43 Morris vehicles (all British) ("Increasing Market" 1934; "New Motor Vehicles" 1935; "More Motor-Cars" 1936).

These automobiles arrived on the southern African scene in this period of intense hunting with both gun and camera. The efficient fuel consumption of the Ford Model T and then Model B earned it a reputation as the truck for the steep hills and dusty roads, one that always got the hunter into the bush and out. Then came Chrysler's Jeep and the British-made

Land Rover, two four-wheel drive vehicles that changed road travel for whites venturing into southern Africa's remote borderlands forever (Capstick 1988, 19, 77).

Yet even for those with motor vehicles, indigenous modes of transport were still indispensable. The terrain virtually held both car and driver hostage, the narrow wheels getting easily stuck in the clay soils of the lowveld known for dusty, sandy winters and sinking, muddy summers. To the present there is no bridge at Masenjeni across Rio Save. When the river is in flood, locals rely on the *xikwekwetsu* (indigenous boat) and pontoons; this was the same transport upon which the Land Rover, Model T, or Jeep were carried in the 1930s-60s when the waters were high. One of the hunter-merchants' favorite haunts was the Panzila area of Masenjeni, not least because it was suspected there might be precious stones (diamonds) there (Capstick 1988, 21–22, 48–49, 58–60, 127–129).

Part III: The Gun and Ammunition in African Hands

The Mine as Hunting Ground for Ammunition

Gold mining began in Gauteng (Witwatersrand) in 1886. By then maTshangana had been going on long journeys to Natal to work on sugar plantations and mills, and, from 1867, to the diamond fields of Kimberley near Cape Town. Their purpose for going to work was to obtain things that could enable them to exercise their own modernities, including arming themselves and hunting with guns in addition to technologies they already possessed. MaTshangana earned a reputation for hard work, for their fearlessness to enter the mineshaft and work underground, and for their ability to acclimate well to the hot and humid conditions. At Kimberley, they had long used some of their earnings to buy guns, or alternatively to receive guns for their payment (Mavhunga 2003).

By 1886 when mining began on the Rand, however, gun and ammunition sales to Africans had been banned. Still, the mines remained an important source of raw materials maTshangana and vaShona could use for making ammunition. As they used explosive charges to blast for ore in the shafts and witnessed one dynamite-related accident after the next, African mine laborers learned that dynamite could be useful gunpowder for their guns (Transvaal Chamber 1897, 3; Mavhunga 2003, 201–231). Despite stringent controls, cases of dynamite (blasting gelatin) were still stolen. On October 8, 1910, for example, the Transvaal Chamber issued instructions that all "natives" returning home be thoroughly searched before leaving the mine premises (Transvaal Chamber 1912 and 1914, 10). Still explosives continued to disappear.

A vivid description of how maTshangana in particular *mined* dynamite from the cartridge cases and transported it home comes from the biography of Bvekenya:

Ammunition was always scarce. Every returning migrant worker would try to smuggle home a supply of powder, filched from the mines by means of furtively unraveling fuses somewhere in the dark, thousands of feet down a shaft.

They would secrete this powder about their persons: in the hair, concealed in their clothing, or hidden in hollowed-out cakes of soap. One man managed to fill up a whole calabash with gunpowder. . . . He had a second calabash full of water to sustain him on the journey.

As he tramped along the Rhodesian border he encountered a police patrol, searching the mine labourers for gunpowder. They stopped him, along with his companions. He sat down miserably, waiting to be searched, while his companions were each forced to give up their small secretions of powder.

The policemen were perspiring in the heat.

"Have some water, masters?" asked the man, with a touch of genius.

The policemen accepted readily. He gave them his calabash of water. They drained it. With his heart in his mouth, he offered the second calabash.

"No," said one of the policemen kindly. "Keep that for yourself, you'll need it on this path."

They searched him and found nothing on his person. He picked up his precious calabash and went on along the path. It was a thirsty journey home, but he sang all the way. (Bulpin 1954, 126–128; Mavhunga 2003, 223)

Journeying or itinerancy became a transient workspace in which explosives were stolen (laboring in the mine), smuggled home (the return journey), and transferred from the mines of the Witwatersrand to the African countryside by Africans themselves (through their mobilities). The return was a fruitful harvest/hunt; in the movement within it we see how ordinary people move things around. MaTshangana's sojourns in the mines were imagined through the imperatives of the village in the countryside, which were met by performing *valungu*'s work to fulfill the imperatives of industrial capital. One act—be it drilling, *ukulaicha* (loading ore onto wagons), or serving as *baas boy* (boss's assistant)—simultaneously produced two different products, for the corporation and employer on the one hand, and for the African villager and worker on the other.

Incoming Things-in-Use

Through their purposive wanderings, maTshangana and vaShona brought back many *svexilungwini* (chiTshangana) or *zvechirungu* (chiShona), or "things of the white man." These "things of the city" (*svedolobheni* in chiTshangana, *zvekudhorobha* in chiShona) were being mobilized as resources for the enrichment of *ntumbuluko* or *chivanhu* (culture or tradition). MaTshangana saw these things, whose mobilities from the mines to the villages they initiated for purposes designed entirely by themselves, as (un)raw materials critical to the execution of their own projects. As we just saw, the colonial appeared in their itineraries as the not insurmountable obstacle; after those obstacles were passed, these designers of destinies continued on their way home. These maTshangana were therefore not merely "people as infrastructure" (Simone 2004)—the basic facilities, services, and installations necessary to enable a community or society to function—but creators and users of this infrastructure.

Two questions become important regarding the aftermath of *svexilung-wini*, when these walkers—these hunters of things useful to themselves—arrived home. First, why and under what conditions do some incoming technologies acquire salience in the lives of Africans and others do not? Second, how do we explain this salience from the perspective of Africans as well as the material properties of these technologies?

It would seem that the scarcity of ammunition governed the conduct of the hunt and spurred African innovations. MaTshangana and vaShona fashioned crude guns from "unrifled pipes which fired pebbles, hard marula pips and a variety of other objects; lethal if they hit the target, more lethal if they exploded in the hunter's face" (Bulpin 1954, 217). They made bullets by running lead around pieces of iron manufactured by the local ironsmiths. Chief Mpapa led his male maTshangana subjects in making gunpowder from the solidified urine of *mbíra* (dassies), which they mixed with the ground charcoal of a sulfurous river plant called *mungwakuku*. MaTshangana also made gun caps by "welting pieces of hide and drying them into the shape of the nipples of the guns and when dry the heads of safety matches were [scraped of their explosives and] put into them." Dampness was the biggest problem with this method of peeling off explosive powder from matchstick coatings, loading the caps, and attaching bullets made from homemade iron balls; the ammunition did not always fire.[9]

MaTshangana brought from the mines disused high-tensile wire cables and barbed wire stripped from European farm fences for use as snares, replacing the organic ones of string made of tree fibers. They also bought the wire from European stores. After setting the traps, they smeared them with a medicine called the *buriba*, a brownish moss that grew on the roots of the *shivumbunkanye* tree. Like the *ndzedzena*, *buriba* was believed to have the power to make an animal lose its power of judgement and to become unsuspecting and reckless (Junod 1927: II, 59).

What Type of Gun for What Type of Animal?

It is important to specify the kinds of innovations particular to certain guns at this point. The first was the Martini-Henry rifle, favored in the period 1890 to the 1930s when its ammunition was still available. In the border-lands, Martini-Henrys were obtainable from the Portuguese colonial offi-cials and shop owners at Chipungumbira (Espungabera to the Portuguese), Masenjeni, and Chikwarakwara just across the border, who bartered them for ivory.[10] The muTshangana and muShona hunter found the rifle too penetrating to the point where after taking a shot clean through, the animal kept charging when the hunter expected the bullet to shake it and stop it dead in its tracks. The advantage of the rifle, however, was that the bullets and cartridges could be homemade. Another problem was that the hunter measured the charge, which was often too much, and the recoil often knocked the firer off his feet as the bullet went off. Measuring the right quantity for each target animal was a challenge, and even if one did, the ammunition did not ignite when the weather was cloudy or rainy.

From the 1930s onward, maTshangana and vaShona hunters in Gonar-ezhou preferred the .303 Lee Metford to the Martini-Henry in elephant hunting. For the buffalo hunter, the .303 was most effective when taking a lung shot from 300 meters. The shot was only possible when the animal was running away, at a 30–45° angle so that the bullet went in through the rib cage to perforate the lung. The other technique was to go where the blood was (heart shot) but it was easier to take a lung shot from the side as the animal grazed or fled at an angle. Of course, if there was no choice and he had to get meat at any cost, then the hunter would wait until the animal was close up facing him, then go for the head shot to "switch off the engine" or perhaps hit it in the midriff straight through the heart.[11] There was one arrest in February 1967 of a hunter living near Malipati dip who used a Lee-Enfield MK III bolt-action, magazine-fed repeating rifle.[12]

The third rifle was the .375 Holland & Holland Magnum, by far the elephant and buffalo hunter's gun of choice (Capstick 1988, 5). African hunters liked using the rifle for its superior ability to penetrate the tough bone of the horns and forehead, which the .303 could not. (Later during the 1970s African nationalist guerrillas deployed the .375 to shoot down enemy aircraft). Whether it was the buffalo or the elephant, the decisive

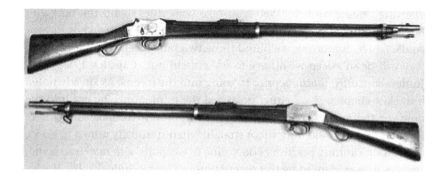

Figure 3.2
The British-made Martini-Henry (top) and its successor, the Lee-Metford, were two of the rifles in use in the Rhodesia-Mozambique borderlands of Gonarezhou.
Source: Author, compiled from Wikimedia Commons

factor was the place or surroundings where the hunter spotted the animal before it spotted him. Wind direction management was key; the hunter would scoop a handful of soil and release the earth slowly to determine the drift and direction of the wind. Whereupon he then approached his target until about twenty-five meters short then "switched off the brain." He avoided at all costs shooting at the trunk because then the bullet would go straight through it, but stop short of lethal impact because of resistance as it tore through muscular flesh. Of course, hunters knew how difficult it was to take the headshot at close range as the elephant charged.[13]

Guns in the Presence of the Spiritual
Having acquired certain *murhi* (medicine) from the *inyanga* or asked for the ancestors to guide the hunt, the hunter could tie down an animal by the

following remote means: "Well, you take your rifle, stamp the butt into the biggest [animal] track, scrape off the dirt, and tie it into a knot in your handkerchief. Tomorrow, we'll find those two bulls sleeping on an anthill. They will be so asleep we'll have to wake them up" (Capstick 1988, 139). Hunters consulted *midzimu* prior to going into the forests. As already noted in the last chapter, *shavé* either possessed the hunter once he got into the forest, or the ancestors led the hunter in the forest to make the correct decisions. A gun could only shoot straight when spiritually armed to do so.

The long-running practice of doctoring of weapons with *muti* (medicine) continued, as evidenced by this description in Charles Bullock's *The Mashonas* in 1927:

The knowledge that a gun will kill better if its muzzle is rubbed in a wound made by a successful shot, is useful [except] only to white men, and they don't seem to take the trouble to do it, even when told. A white man would not even believe that it is necessary to doctor a gun after shooting *mombo*, the bad old dog baboon who led the herd. There is a medicine the smell of which no elephant can resist. It was sprinkled on the ground, and, near by, the hunters lay hidden. He danced in front of them, and they, being attracted by his white dress, followed him until he led them to a place, where they could smell the sweet scent. They stopped to dig with their tusks, and the hunters sprang out with assegais, bows and axes. (Bullock 1927, 255–256)

We have already seen medicines at work in the context of *ndzedzena*; guns too were believed incapable of killing unless they were doctored and given spiritual armament. The powers that their designers overseas might have given such artifacts to "configure the user" (Woolgar 1991) are not in evidence. Instead, vaShona and maTshangana are configuring the artifact in such a way that its material properties alone are worthless to execute the task of killing an animal. Those material properties are not being denied; they are simply not adequate. A value-addition process is taking place, a synthesis, not of faith and reason, but faith as reason, reason as faith—faithful reasoning as much as reasonable faith.

Guns as "One Other" Repertoire among Many

As already noted, by the 1870s, maTshangana and vaShona were deliberately traveling with the objective of bringing back guns and slotting them into roles that added value to the repertoires already in their possession.

The best examples of this are tracking and stalking. Ammunition was scarce, taking rash shots an unforgivable sin. Every shot had to count, such that one spent excruciatingly long periods of time waiting for an animal "until (it) eventually came so near that it was next to impossible to miss and (the hunter) could use the barest minimum of powder" (Bulpin 1954, 212). This frugal approach to ammunition use was an extension to the gun technologies of indigenous hunting skills of stalking. Such economy had prompted the English hunter-artist John Millais in 1895 to remark that maTshangana hunters had elevated "the noble art of spooring" into "a science" better left to the "black companion" (Millais 1895, 117).

Another British traveler, Parker Gillmore, refused to agree with Selous that the "true measure" of African hunting skills was "shooting straight." For Gillmore, everything boiled down to producing the shooting opportunity to start with. Selous had been astounded at "what bad shooting [African hunters] made; their bullets kept continually striking up the ground all round the elephants, sometimes in front of their trunks" (Selous 1881, 86). But that was because he insisted that the hunters he contracted must shoot from afar, according to the technical material specifications of effective rifle range. Gillmore observed that such factory range was irrelevant to most African hunters, who "never fired at a beast at a distance of over ten or twelve yards, and accordingly made sure of [their] shot every time" (Gillmore 1890, 118–119). MaTshangana hunters told Millais as much in 1893: hunting on their own, they used their bravery, endurance, and stealthy encroachment skills to creep close enough to make every shot count. But now their white employer always ordered them to shoot from afar. They missed. It was his fault, not theirs (Millais 1895, 117).

To avoid wasting scarce ammunition, maTshangana hunters deferred to and adjusted a number of their indigenous hunting methods. When going into the forest, the hunter carried his gun, bow and quiver of arrows, a knife, and an axe.[14] In the forest hunters made poison from *vutsulu* to tip their arrows and pellet gun bullets to kill kudu and other antelope at close range. Such pellet guns would have originally been designed for killing birds at a range of fifty meters. The stricken animal would run for a while and then fall dead inside thirty minutes, but the meat was not poisoned.

The symbiotic relationship between *valungu* and some local hunters relied on the knowledge that the white man was wasteful: he killed plenty of elephant only for the tusks, leaving the giant carcass to rot. So when people heard a shot or saw vultures circling, they headed in that direction knowing that a white hunter had just felled an elephant or one was in distress and soon would die. Aware that the shot ringing in the air would have advertised that he needed free labor to chop out the ivory, the white hunter would ordinarily fire another shot as further guidance to his

position. The journeys to the elephant carcass that we have already seen—
from predators, birds of prey and carrion, to villagers—began in earnest
(Bulpin 1954, 55, 65).

Conclusion

This chapter has shown how the professoriate of the hunt provided an
anchoring structure through which incoming things like guns found
salience among maTshangana, and vaShona. Guns are an example of how
Africans throughout their histories have assembled around their indigenous innovations the bits and pieces of raw materials—human, machine,
finance, and even countries—from all over the world, and through trading
marshaled these components into instruments for achieving their own
ends (Bayart 2000, 217). It has been shown that this tendency of Africans
to look outward reflects capacities to identify gaps within their repertoires
that incoming things might address. As such, the global circulation of
technology and indeed its definitions and purposes, can never be fully
understood unless we include people who do not create the technology,
but who make it work for them in a certain way according to their own
objectives. No matter their powers of invention, designers in the conventional (Western) built laboratory cannot capture in their design the totality
of life-worlds that would-be users inhabit and within which their technologies will be used. Nor are these designers' artifacts immune to such agencies
once "out there."

Earlier I explored this latitude to dictate the fate of the incoming things
in a context where the incoming people purveying them have become
colonizers. I suggested that, instead of rupture, the patterns of creativity
and initiative that we have seen in this chapter continued into the twentieth century, in particular because colonial rule was virtually absent on
the margins.

This chapter then unsettles the meanings of technology and renders
them contingent. Is technology the guns? What guns—the pipes Africans
assembled and fired or the .303s whose specific efficacy depended so
completely on the *unyanzvi* or *vutivi* of the hunters? The bullets? Which
ones—the marula pips, pieces of metal cut and shaped by ironsmiths-
turned-gunsmiths or those traded from *valungu* whose hunting forays
would have been nonexistent outside of local kinship and the professoriates of the hunt? The gunpowder? Which one—from the droppings of
dassies and other niter made locally or the dynamite fleeced from the

mines and smuggled in calabashes through the dragnets of police patrols?

Throughout the existence of Rhodesia (1890–1980), the immediate border area demarcated by the Save to the north and the Limpopo to the south remained a vast margin untouched by colonial rule, where the professoriate of the hunt flourished as a result, virtually without the interference of the colonial center. The animals were there; rinderpest (cattle plague) and settler appetite for hunting for the pot had cleaned the settled hinterland of animals but not the borderlands. The guns were there; here Europeans could leave and return from colonial cities with supplies of the latest guns and ammunition aplenty. The market, despite the cartels, was there. The state was not.

Conventionally, when scholars talk about "the colonial period" as one of European settler or imperial power, they make it seem as if every particle of soil was occupied by a colonial regime and every African inhabitant all of a sudden felt its grip. This assumption of colonial power and its "isms" and "schisms" does not apply to the margin where the use of guns in hunting and the ivory trade continued to perpetuate the same kinds of guided mobilities prevailing before the partition. The more important question to ask about guns on the margin is not how the colonial state used them to assert its power, which is true of the hinterland especially in 1896–1897, but rather, how Africans weaponized the fact of their distance and remoteness from the hinterland to help themselves to the bounties of nature. In an important contribution, Jane Guyer (2004) has called such innovations on and because of location within borderlands "marginal gains."

This chapter has examined the absence of the state—with all its signifiers and enforcers—owing to the distance of this place from state capitals like Lourenço Marques, Salisbury, and Pretoria. The face of the state—police, army, district administrators, clearly marked boundaries, and the rule of (government's) law—was visible only by its near-complete absence. Temporal bifurcations of historical time, such as precolonial, colonial, and postcolonial, exaggerate rupture at the expense of persistence. Spatial bifurcations reify "national histories" as if people on the ground respected—let alone saw—the borders of "Rhodesia," "South Africa," or "Mozambique."

Therefore, the spatiotemporal demarcations and the powers often accorded to colonial rule are all based on what was happening at the center, not the periphery. There was no clean break between the precolonial, colonial, and postcolonial, just as a hunter in Gonarezhou today might be a mineworker in Johannesburg tomorrow, or staying with relatives in South Africa where, apparently, he might also be married to another wife, with children, and fields to plow, and where he has another homestead and is part of the community (Mavhunga 2007a). Ultimately, the republic of absence ushers in a republic of mobility, whereby the most powerful technology for exercising one's democratic choices is the ability to move and make through moving.

4   Tsetse Invasions

The Prime Minister of Cape Colony, Cecil John Rhodes, was a very ambitious British capitalist. After establishing the British South Africa (BSA) Company, incorporating it under royal charter in 1889, and floating its shares on the London Stock Exchange (LSE) Rhodes set out to colonize lands to the north of the Limpopo River which separates present-day South Africa and Zimbabwe. In 1889, by deceitful means (the fraudulent Rudd Concession with King Lobengula of the Ndebele), Rhodes raised an invasion force of three hundred men to occupy the lands of vaShona (hence the British corruption "Mashonaland"), each man buoyed by promises of land concessions.

In 1890 the volunteers of the Pioneer Column (as the invasion force came to be known) marched into "Mashonaland" and settled down to prospecting and mining. Hopes of finding another Rand (gold deposits like those in South Africa) were sky high. Some (but not all) vaShona people indeed welcomed the Company, not as masters, but as allies or sources of guns against their local rivals and the notorious raids of the *madzviti* (Beach 1974; Lipschultz and Rasmussen 1989, 125). VaShona used the term *madzviti* to refer to two states notorious for their devastating raids in search of cattle, slaves, and tribute, the Ndebele to the west and maTshangana to the east.

The pacification of African resistance and the containment of a rinderpest outbreak (happening simultaneously in 1896–1897) gave the Company some breathing space in two respects. First, the combination of military defeat and loss of cattle (for draft power, food, and wealth) drove Africans to seek work for meager wages on the emerging European settler farms and mines. Second, the rinderpest's extensive devastation of livestock and forest animals also exterminated the tsetse fly parasite in many areas, opening them up to European settlement (Palmer 1977, 40, 42, 51, 93).

By dismantling African spiritual and political sovereignties over the forest through a combination of land expropriation and pathogens like rinderpest, the colonial settlers may not have known that they were rendering themselves vulnerable to the scourge of tsetse fly. When the animal populations began recovering and started pushing out from the forested border regions toward their former haunts, the settlers called this an "invasion" and an "advance," and erected "tsetse fronts" against it.

Here was an insect that did not travel far on its own wings, and relied, therefore, on the mobilities of other organisms (principally forest animals like elephants and buffalo) and those of humans and their technologies of transportation. Thus the tsetse invaded uninfected places by lurching onto walking, biking, driving people who might be visiting friends, running government errands, and hunting. These animal species became organic vehicles for the parasitic passenger; which turned their bodies into what Thrift (2004) called "movement-spaces" or mobile physical places upon which the insect could sit, cling, and feed. The insect-passenger turned the itinerary of its carrier into pestiferous mobilities; the innocuous intentions and intentionality of its human and animal carrier supplied the insect with long-distance invasive power.

Together, the organic vehicle and its passenger become intertwined as vectors of the trypanosome; the outbreak of trypanosomiasis becomes co-produced "through a conjunction of bodies, technologies, and cultural practices" (Sheller 2011a, 6; Sheller 2011b). Here is an example where an insect subverted and indeed mediated an entire colonial project and inserted itself as an indefatigable pest. What then is to be said of the much-vaunted powers of imperialism and colonialism when those who are supposed to be the lord of the African become hostages to arthropoda?

Revisiting Ecological Imperialism

The colonial historiography depicts a physical environment in peril that European technology pulled from the brink. A critique of this view has been given (Beinart 2000). A contrary view sees a symbiotic relationship between people and forests that colonialism reversed (Kjekshus 1996; Fairhead and Leach 1996; Ranger 1999). Others say the precolonial people-forest relation was not perfect yet controlled, and that once colonized, Africans watched helplessly as anti-tsetse fly "game slaughter" and vegetation clearance opened up tsetse and mosquito habitats to *development* (MacKenzie 1988; Mutwira 1989).

Africans were displaced to overcrowded "native reserves" as their erst-
while resource-rich lands became game reserves, white-owned farms, mines,
plantations, and towns (Guelke and Shell 1992). New regimes of forest,
animals of the forest, soil, and water regulation went simultaneously with
campaigns against tropical diseases (Beinart 1984; Anderson and Grove
1987). Western European cultures of environmental science began to direct
health and capitalist development (Anderson 1992, 508). A clear separation
between fenced-in nature and fenced-out culture nullified and criminalized
the spiritual views and access of Africans to the sacred forest. The only
sacredness now was the law.

Many scholars have noted how colonists used "environmental control"
and "development" as cover to seize *ownership* (not just *control*) of natural
resources from Africans. The science-anchored condemnation of "destructive" spiritually guided mobilities and practices within the environment
became the basis for dispossessing and alienating Africans from lands they
had long domesticated and forests that had thrived because they were
sacred (Anderson 1984; Moore and Vaughan 1994; Ranger 1999; Driver
1999; Peters 1994). Where these forests and their teeming herds were
found, such "ecosystems" were attributed to the pristine: they had not yet
been touched by the "destructiveness" of the African. In the case of
South Africa and Rhodesia, able custodianship of nature became a racial-
ized question: aesthetics toward nature was a white man's natural disposi-
tion (Carruthers 1995; Wolmer 2007). The idea of protected areas and
ecotourism, both stimulated and emerged from this dismissal of "the native
as arch-destroyer of nature and the white man its knight in glittering
armour" (cf Rajan 1998; Cronon 1983; Ranger 1999). With colonial bureaucrats and Western science in charge, Africans were compelled to abandon
age-old crop, animal, and soil practices and follow defective Western
science amid flogging, sweat, and labor (Drinkwater 1989; Palmer 1990).

The sections that follow build on an argument that Helge Kjekshus
(1977) made regarding the disruptive impact of European imperialism
upon people-forest relations, contrary to the colonial project's self-portrai-
ture as technological redemption to a precarious ecological situation. In a
follow-up, Kjekshus (1996) showed how East Africans had developed "agro-
pastoral prophylaxis" involving, inter alia, setting fire to well known tsetse-
infested areas every year just before the onset of the rains, and avoiding
pasturing livestock in them, as Ford (1971, xxx, 474) had noted. In their
movement into and settlement within east Africa, Europeans brought
with them three biological agents that changed everything: the rinderpest,

smallpox, and jigger fleas. Rinderpest "broke the backbone of many of the most prosperous and advanced communities, undermined established authority and status structures, and altered the political contacts between the peoples. It initiated the breakdown of a long-established ecological balance and placed nature again at an advantage." For that reason, Kjekshus reckons that rinderpest "represents the dividing line between initiative and apathy on the part of a large number of African peoples, particularly in the eastern and southern parts of the continent" (Kjekshus 1996, 127).

This story is not peculiar to Africa. The American historian Alfred Crosby also attributed what he called "neo-Europes" (colonies modeled on European lines) to bio-ecological processes, specifically carriers of pathogens that accompanied or were passengers in their bodies. Europeans carried to the Americas and Australia pathogens to which they had gained natural resistance through exposure to, devastation by, and recovery from smallpox and measles (in humans) and rinderpest (cattle). The opening up of these new lands brought indigenous people face to face with a most hideous enemy: "not the white man . . . but the invisible killers which those men brought in their blood and breath." He showed that imperialism was accomplished not through technological means (alone), the sort of *technological imperialism* that Headrick (1981 and 2010) talk about, but through organisms that changed the ecology entirely, what Crosby (1993, 31) called *ecological imperialism*. I argue that these pestiferous mobilities necessitated the colonizer's recourse to the professoriate of the hunt—among other indigenous innovations — for a solution.

The Rinderpest as Agent of Ecological Imperialism in Zimbabwe

Indeed, scholarship on Zimbabwe has explored the effects of the rinderpest on cattle and how they both triggered the 1896–1897 risings and determined vaShona and Ndebele's responses to them (van Onselen 1972). Similar effects on people and their livestock have been documented in the case of South Africa and Lesotho (Phoofolo 1993 and 2003). Of late, the spread of the rinderpest has been accounted for, as a kind of journey from southern Europe via Somalia, Egypt, and the Sudan, through Kenya, Malawi, and Zambia into Zimbabwe, until it was stopped at the gates of Cape Colony (Spinage 2003). The scientific aspects of the panzootic, along with the institutions and physical infrastructures of veterinary science that emerged in southern Africa, particularly in South Africa, have recently attracted belated but distinguished interest from historians (Brown and Gilfoyle 2007).

My interest therefore lies in other gaps, specifically how the rinderpest almost exterminated the vectors and hosts of tsetse fly—big forest animals—from much of Southern Rhodesia barring the margins. For while rinderpest also devastated livestock, it did not spare the forest animals, especially those that tsetse fly also relied upon for its blood food. Hence the rinderpest left a tsetse-free environment enabling the cattle ranching industry to thrive while limiting the tsetse's habitat to the margins of Rhodesia. This discussion on how the rinderpest that the Europeans brought devastated the tsetse fly sets us up for a "counter-ecological imperialism": by insects pushing back and recolonizing the lands from which they had been earlier displaced.

The rinderpest arrived on the Zambezi River on its long journey from the north as a passenger (if an unwitting one) in the fluid bodies of oxen drawing the wagons of Europeans. In a massive 765 pages, Clive Spinage has chronicled the ultimate rinderpest story, so comprehensive as to make any future attempts at addition look amateurish. He shows clearly how the strain came into Africa from southern Europe as a microscopic passenger inside the bodies of live Russian cattle that landed in Egypt in 1841. Afterward it spread through organic mobilities—of domestic and forest animals and people—and in particular wagon and pack animals, right up to the borders of Cape Colony, where it was finally halted by recourse to inoculation in 1897–1998 (Spinage 2003, 497–549).

Under normal circumstances, when Africans were still in charge of their own lands, these infected wagons would have been stopped right at the border post of the chiefdom, kingdom, or empire. Take, for example, how Mzilikazi ka Matshobane, king of Ndebele, had given instructions to halt European missionaries about to bring lungsickness-infected cattle into his domains in 1859. The original account is told by the missionary Robert Moffat in his detailed diary, but the historian Ngwabi Bhebe summarizes it rather well:

On approaching Ndebeleland, the missionaries observed signs and symptoms of lungsickness among their draught oxen and immediately reported these to Mzilikazi. The king not only ordered the quarantining of the missionaries themselves. An *inyanga* (healer) cleansed them and their possessions by sprinkling them with medicines soaked in water. Apparently the missionaries were ignorant of the whole significance of the cleansing ritual. One of them, William Sykes, had lost his wife at Kuruman and this was known to the Ndebele. To the Ndebele Sykes was carrying on his person the evil and contagious force of death which was already affecting the draught oxen and should he be allowed to come into the country untreated he might affect the whole nation together with all their livestock. During and after

such a treatment, the bereaved person as well as his surviving relatives were supposed to stay away from the rest of society. In fact for almost three months Mzilikazi
and his people treated their missionary guests as outcasts, refusing even to show
them a place for settlement. (Bhebe 1979, 28)

The new colonial occupiers only observed quarantine measures too little
too late to stop the rinderpest from causing the transport system—then on
the hoof—to collapse. As Pule Phoofolo shows, the pest would have swept
south past the Zambezi River even sooner, but the giant river delayed that
journey "probably due to the lack of an intensive network of cattle communication across the river" (Phoofolo 2003, 504). Once it breached the
barrier in early 1896, it hitched a ride on ox-wagon traffic and forest
animals, and cruised south. Subsequently, its merciless slaughter of oxen,
the linchpin of the transport system, brought the transport sector down
on its knees, with grave economic consequences.

There are significant merits in exploring the rinderpest migration as
ecological imperialism or as pestiferous mobilities. We see for example the
role of transport systems in conveying the pest, turning wagon roads into
linear dispersal points for the microbes. In particular, the waterholes at
which the wagons stopped to allow oxen to drink, the campsites where the
wagons outspanned at sunset, and pastures where the cattle were let to
wander around and to graze at night, all became sources of contagion.
Charles van Onselen (1972, 473) was probably right that "the Zambezi was
possibly the most effective barrier which the rinderpest encountered in its
travel southward." As he said, the "relatively well developed communications" and "the extensive use of the ox-wagon" speeded up the disease to
"compensate for its protracted stay in Northern Rhodesia."

By pointing toward ox-wagon traffic, van Onselen leaves room for us to
search for the "breech points" on the Zambezi as the likely bridges through
which this "germ traffic" passed into Southern Rhodesia. However, we
might also overlook other *nzira dzakahwanda* (hidden pathways) along
borders. Archives indicate that the government's understanding of the
rinderpest's source was that cattle from Ndebeleland (hence European travelers' distorted name 'Matabeleland') had come into contact with forest
animals in the Zambezi valley and spread the *mukondombera* (epizootic) to
Bulawayo, where it spun out of control as the oxen drawing the wagons
caught on (Montagu 1896, 1121; Hutcheon 1896, 1027). However, it is
open to debate whether, in fact, it was wagon traffic, local inhabitants visiting relatives, or animal movement—including carrion-eaters—that spread
rinderpest.

It cannot be discounted that rinderpest might have spread through the upper Zambezi wild animal belt direct into the Chobe area and thence carried into Palapye. Besides authorities in Bulawayo district noticing "buffaloes and large game dying in immense numbers," it was also clear that cattle from north of the Zambezi were bringing the disease into Ndebele territory. As the Native Commissioner for Gwanda, Val Gielgud observed in 1908, rinderpest swept to death seventy-five percent of the cattle in Ndebeleland on its way south, while in northern areas where animals were plentiful, its concentration on the few available waterholes wrought "great havoc." Gielgud observed that in well-watered areas, rinderpest caused less damage than in arid areas (Gielgud 1908, 35). Ndebeleland is generally dry country.

From a transport history perspective, the larger story of the rinderpest is that it hastened the process of railway construction in southern Africa, since draft oxen had been devastated (Pirie 1993b). Africans could still walk long distances, but the mines and farms needed a bulk-carrier type of transport capable of moving produce and raw materials over longer distances. In the wake of the plague, few bullock wagon spans remained on the Cape Colony to Bulawayo/Salisbury route. A revolution in transport systems was underway as mule and donkey replaced scarce oxen; more importantly, the acute transport shortage drove the construction of the Mafeking-Bulawayo railway line into overdrive. African laborers built this railroad, as they did the roads, not only in Southern Rhodesia but elsewhere in the region as well (Zhao 1994; Chiteji 1979; Chilundo 1995; Sheldon 2002).

By eradicating the tsetse fly in its pathway, the rinderpest opened up for Europeans the possibility not only of reintroducing draft oxen, but also of enabling human settlement, albeit temporarily given that the tsetse fly would reinvade its old habitats by way of expanding animal populations searching for greener pastures and water. As rinderpest devastated bovines, it eliminated most of the tsetse fly's blood diet, hosts, and "transport." As the tsetse died, it became possible to replace the readily "vanquishable" draft ox with the donkey, horse, and mule—themselves fairly resilient toward rinderpest ("Results of Rinderpest" 1897, 459).

The large numbers of settler caravans passing "almost daily" through Bechuanaland represented "substitutes for the old methods of transport riding" that went apace with the plague. This scarcity-induced technological innovation did not end with the replacement of "ox-power" with "donkey-/mule-power" but also hastened the introduction of trains, and

later automobiles, in southern Rhodesia discussed earlier ("Results of Rinderpest" 1897, 459).

The oxen scarcity created a market for overseas transport technology.
The "local" stock breeders augmented their horse, mule, and donkey stock
through imports from the UK. So thick was the commercial intercourse
that by mid-1897, "nearly every steamer sailing for South Africa [took] out
a few good pedigree animals for breeding purposes." For example, the
Gascon was scheduled to sail to South Africa on October 1897, carrying
among its cargo "a hackney stallion and a hackney mare as well as two
Cleveland mares, while a Jack donkey [could] be shipped by the *Scot* for
Algoa Bay" ("Results of Rinderpest" 1897, 459). It makes one wonder
whether the many donkeys in Ndebeleland and Botswana today are, in
fact, progeny of these imported post-rinderpest breeds.

The Invading Tsetse: Or, Reverse Ecological Imperialism

In the colony of Southern Rhodesia, the effects of the rinderpest were not
uniform. In the core areas of European settlement and traffic, where ox-
drawn wagons passed through, the impact was severe. On the margins of
the colonies, places that were inaccessible to ox-wagons, traversable only
on foot, by birds in flight, or forest animals on the hoof, the rinderpest's
effects were mild to nonexistent. It was from here that the tsetse invasion,
a *reverse ecological imperialism* of sorts, began.

By this I mean the tsetse and their forest animal hosts were merely
returning to haunts from which the rinderpest had displaced them. By
these pestiferous mobilities, the little insect threatened all the economic
infrastructures (principally the cattle ranches) the settlers had installed in
the belief that rinderpest had exterminated this carrier of trypanosomes.

The discussion of the tsetse fly invasion must start with the subtle
dynamics of rinderpest-animal-tsetse interactions and the effects on the
margins of Rhodesia. For the sake of clarity, focus will be limited for now
to the southeastern lowveld area bordering Mozambique before extending
to the whole colony later in the chapter. In the Muzvirizvi (which the
Portuguese corrupted into *Mossurize*) area of Mozambique abutting the
Chipinge district of Southern Rhodesia, local maTshangana and maNdau
inhabitants had told the entomologist Charles Swynnerton in 1921 that
the effects of the rinderpest had been quite mild and shortlived. West and
east of the Sitatonga Hills (in the same area), the tsetse fly had continued
its life as if nothing had happened. Even when rinderpest destroyed the
insect's food source (wild ungulates), the tsetse fly had simply switched its

diet to human, bushpig, warthog, and baboon blood, pouncing upon these whenever they made themselves vulnerable.

North of the Rusiti (Lucité) River that runs through Muzvirizvi, prior to rinderpest there had been plenty of buffalo, less so in the south. Both areas were insect-infested. North of the river the rinderpest had resulted in reduced tsetse fly for two straight seasons. By the third, the tsetse had fully recovered, even if buffalo was still scarce. It seemed that "the fly had previously become especially dependent on the buffalo, owing to the latter's immense relative numbers, and took a season or two to adapt itself completely to the habits of the other larger mammals" such as warthogs and baboons. South of the river, in the fly-infested areas east and west of the Sitatonga Hills, "no noticeable reduction of the fly" followed the rinderpest. However, the insect "bothered [people] terribly"; one local muTshangana told Swynnerton that "perhaps with their other food destroyed [tsetse] attacked men more, so that, though fewer, they seemed as many as ever." Most forest animals had died but bushpigs, warthogs, and baboons had lived (Swynnerton 1921, 343).

Swynnerton's conclusion from Muzvirizvi helps put later characterizations of the tsetse fly as invading Rhodesia from "colonies" inside Mozambique into perspective: "The actual effect of the rinderpest in certain areas," he says, may have been not directly to starve the fly, but "to confine it the year through to a far greater extent to its dry season centres by the destruction of its chief carriers" (Swynnerton 1921, 344). The description of the tsetse movements in the colonial government correspondences as *invasions* is critical because, as we will see in the next sections, such militaristic language was extended to designate outbreak areas as "tsetse fly fronts" or "defense lines," places where combat against the insect was to be mounted.

It is necessary to explain what the Branch of Tsetse and Trypanosomiasis Control (whose name changed several times) meant by tsetse "invading" and "advancing." The insect caught rides on anything that moved (hence the term "carried fly"). It was a passenger on mobile forest animals and domestic animals (its food source), and mobile people, who carried it on their persons, bicycles, and motor vehicles from tsetse-infested areas to tsetse-free areas. The insect, the human, the forest animal, the car—all these became self-propelled or *auto-mobile* bodies moving in ways that conveyed the trypanosome to livestock and humans. Tsetse flies and forest animals in particular were in this sense *organic vehicles* carrying trypanosomes from known infected forest animal populations to uninfected cattle pastures.

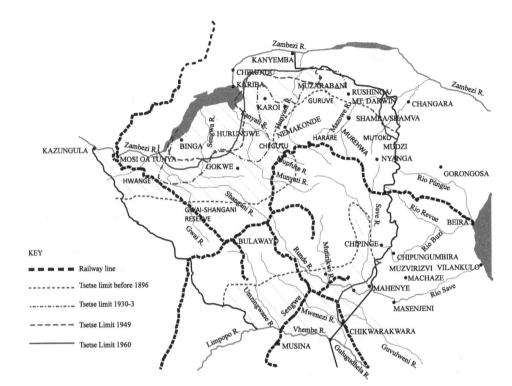

Figure 4.1
The limits of tsetse fly and the areas that became "tsetse fronts."
Source: Author

By 1909, the first serious signs of a return of tsetse to its pre-rinderpest haunts appeared. The few animals that had survived the plague in the colonial hinterland had been all but hunted out. Most of those that remained were to be found far from areas of concentrated human occupation—principally along and east of Southern Rhodesia's border with Portuguese East Africa (Mozambique) and along the Zambezi River. As their numbers grew due to natural increase, the animals began ranging outward, with tsetse fly in tow. At the same time, to cope with the colonial challenges to livelihood, and to escape from—and find money for—paying taxes, Africans were traveling across tsetse-infested lands to seek work in the mines of central Rhodesia and South Africa. White farmers, Native Commissioners, and illicit labor recruiters were raiding Africans for taxes and labor.[1]

Ecological imperialism, therefore, was a kind of *spatial disarmament*, or the dispossession of people from spaces they had settled in and modified over time through experimenting and consolidating what was now known, in order to control nature or human enemies. The European occupation and its resultant movements disrupted the proper functioning of prevailing African anti-tsetse stratagems and infrastructures. The high rates of trypanosomiasis infection in cattle demonstrated the folly of moving livestock through tsetse-infested areas by day and proved just why Africans had moved by night when cold temperatures rendered the insect immobile in precolonial times (Anderson 1856, 488–489; Kirk 1865, 154; Baines 1877, 65–66).

Later on, the tsetse Branch would use tsetse's behavior in certain temperatures in exposing or detecting the presence of the insect by simply walking a black ox through the bush, knowing tsetse would lurch onto it and engorge its proboscus. Indeed it bears stressing that the tsetse was capitalizing on transgressive mobilities arising out of the breakdown of restrictions of access to and passage through spaces hitherto avoided. Places considered sacred and to which access was possible only through the observance of taboos were now passed through and settled in. Cities became new magnets for settlements, eliciting movement across diseased spaces. People were dispossessed from vantage points from which they had controlled problematic nature, and forcibly resettled in areas where they were controlled by it.

The removal of prevailing discipline on movement and the beginnings of new pathways disarmed past ecological philosophies and spatial armings of nature against pests. All these mobilities attracted tsetse fly, an insect always alert to anything that moved during the daylight; it wasted no time catching a ride and traveling as carried fly to new places, where it alighted and, on feeding upon new hosts, shared the protozoan gifts it had carried. Throughout the 1910s-60s any spotting of the insect itself or outbreaks of trypanosomiasis could only mean one thing: a tsetse "invasion" or "advance" was underway.

The brief of the sections that follow is to undertake a discussion of the areas of tsetse fly invasion or advance. As will become immediately apparent, the tsetse fly "colonies" from which the tsetse advanced into the Rhodesian hinterland were located in the areas that the insect had traditionally occupied prior to colonial partition. Indeed, the tsetse fly was simply coming home to roost.

The Northern Advances

The first tsetse advance took place in Rhodesia's northwestern district of Sebungwe. Before 1896, tsetse in Wankie District extended as far south as latitude 19° and covered half of what later became Wankie (Hwange) National Park.[2] After rinderpest, only a few colonies were left. But the insect recovered rapidly.[3] In 1909, after increasing alarm at its resurgence, the BSA Company administration of Southern Rhodesia appointed a chief entomologist to see to it that tsetse fly, deadly to both people and livestock, was crushed. In 1918, the fly was discovered to be advancing into the Gwaai and Shangani Native Reserves from the Zambezi (figure 4.1). European farmers appealed for government intervention.[4]

Hunting operations started in Sebungwe in 1919 to kill big game animal hosts of tsetse and were so successful that they were extended to the entire fly-infested countryside and beyond Sebungwe to Darwin (Rushinga), Hurungwe, Kariba, Chirundu, Chegutu, Kadoma, Sipolilo (Chipuriro or Guruve), Lomagundi (Nemakonde), and Mutoko (Mtoko) districts.[5] The Sebungwe tsetse belt remained active for the entire period of white settler rule.

About the same time that tsetse fly was being reported in Sebungwe, another outbreak was developing in the Chegutu area. Following the start of mining operations, clearance of vegetation in preparation for the mines and miners hunting for the pot exterminated forest animals locally and alongside them the tsetse fly. In 1926, the insect started "threatening" the white-owned farms in the Golden Valley area of Kadoma district. Fences were strung up thirty miles long and ten miles apart. In these corridors, shooting operations commenced to create "game-free cattle-free" areas separating infected from "clean" areas. The operations achieved such success that they were extended west into Gokwe-Sanyati and north into Nemakonde and Guruve.[6] The Chegutu and Kadoma fly outbreak was snuffed out but skirmishes elsewhere continued throughout the colonial era. The Gokwe-Sanyati tsetse belt abutted another in the Kariba-Hurungwe area to the east-northeast, where operations were commenced in 1942.[7]

To the northeast, a tsetse invasion from Mozambique was reported in the Kandeya Native Reserve of Darwin (Rushinga) in 1923. By 1946 the frontage of invasion had widened to the neighboring Mutoko district adjacent to the Gorongosa fly belt in Mozambique. The first flies were caught inside Rhodesia in 1950, as the infection escalated in Chimanda, Chiswiti, and Muzarabani areas of Darwin. Tsetse management officials worried that "the tsetse on the Mozambique side may be quite close, if not over, our borders." The international boundary was now defined very much as "a defence line against tsetse."[8] From 1965 to 1968, outbreaks of trypanosomiasis ran the northeastern areas ragged, devastating cattle herds in the

Katsande, Kapondoro, Mutedza, and Marira areas of Darwin, while minor outbreaks hit Chikore, Magondo, and Sarakurima.[9]

By 1956 the tsetse had swept south into Nyanga North. Previously the fly was known to exist in isolated colonies in Mozambique "a few miles from the border." Cases of trypanosomiasis had taken place on the Ruenya River, but no thorough tsetse survey or operations had been carried out. The Tsetse Branch's verdict was unambiguous: "The position may remain static or grow progressively worse. If the position deteriorates then game elimination operations will have to be commenced."[10] By 1960, the build-up on the Kaerezi River right on the Mozambique border had become "thick." Bush-clearing operations were undertaken from 1963 to 1965 with minimal impact.[11] Surveys in the area between Ruenya and Kaerezi drainages in 1966 showed concentrations of *G. morsitans* in the neighborhood of Chimusasa, Chapatarongo, and Mutarazi.[12]

From Nyanga going south, the only other significant fly front was in Chipinga (Chipinge), from the Sitatonga area, south to the Lundi River.

The Chipinge Advance

Long before 1896, the conditions for the post-rinderpest tsetse fly resurgence were already being born. We have already met Gaza king Manukusa Soshangane's son Mzila returning to seize the throne from his brother Mawewe ka Soshangane. Mzila's son and successor Ngungunyane had set aside his father's targeted animal elimination, vegetation clearance, population resettlement, and rigid policing of forest animal-livestock boundaries, choosing instead to protect forest animals to cash in on an increasingly lucrative hunting safari market. Said maTshangana chroniclers whom Swynnerton reduced to "informants":

[Ngungunyane] had decreed *azi-zale* (let them multiply), and game had become more abundant both outside and inside the cattle-keeping areas. The guard-areas still opposed its passage into and out of these areas and no harm resulted to the cattle. When the population left, the game . . . just "burst forth" (*za-dabuka*). At the same time the wooding was let loose and soon re-established itself throughout the previously settled country. . . . In a very few years [by 1896] the fly had more or less regained its old wet season limit—not many miles east of the present political boundary. (Swynnerton 1921, 335)

It was in 1889 that Ngungunyane and his maTshangana followers left the Chipinge highlands and trekked to Bilene near Lourenço Marques to escape tightening British and Portuguese encirclement. Four years after

Ngungunyane's departure, the "Gazaland trekkers"—or a group of *mabhunu* (chiShona for "white man/farmer") and their families—arrived in Chipinge and renamed it Melsetter. Tsetse fly greeted them cheerfully, with sporadic trypanosomiasis cases in cattle occurring on the highlands adjoining the Mozambican border.

When the rinderpest wiped out the tsetse fly in 1896, residual colonies had remained in parts of Mozambique shielded by the mountain ranges stretching from Chipinge all the way to Nyanga. During the 1910s and 1920s, these scattered but expanding colonies gradually coalesced into larger tsetse belts.[13] The Zambezi salient experienced this tsetse recovery much faster than the eastern borderlands, probably because of the favorable (hot) temperatures, low-lying altitude, and warm moisture in the Zambezi soils in contrast with the colder, mountainous eastern highlands. But in 1915 a serious outbreak of trypanosomiasis was recorded.[14] In 1918, Swynnerton found the southernmost limit of the fly belt to be 20° 20′ (33° 25′ W), on the Buzi river area of Mozambique to the west of the Sitatonga Hills. Three years later, he warned of the "disquieting fact that the tsetse *Glossina morsitans* is slowly spreading west through the lowveld towards our border," and there was now a distinct threat that the fly might in the future invade the Save River valley ("Annual Report" 1921, 315).

Things turned for the worse in the 1930s. In 1932, the Mozambican areas from the Chimanimani Mountains in Chipinge to Jersey Farm (Beacon 96) came under the spell of four tsetse (*Glossina*) species: *G. brevipalpis*, *G. pallidipes*, *G. morsitans*, and *G. austeni*. The first three had been apprehended in Southern Rhodesia; *austeni* had been picked up on the Rusiti River very close to the border by Portuguese entomologists. In the stretch from Mayfield Farm to Jersey Farm, the government had embarked on border forest clearing, which was "extended and maintained by slashing and burning each year" for the next two decades.[15]

The area comprised land of varied altitudes ranging from 6,000 feet at Chipinge to 3,800 feet at *gomo re Chirinda* (Mt. Selinda), followed by the Rusiti and Buzi Rivers, which crossed the border at below 1,500 ft. altitude. This well-watered and fertile country was host to European farms, government forest areas, tea estates, the Ngorima Native Reserve, the Tamandayi and Gwenzi Native Purchase Area, and the Melsetter Native Purchase Area No. 6. The border clearings had been "reasonably successful" in halting the tsetse fly advance and controlling trypanosomiasis. Tamandayi worried the Tsetse Branch: "The [African] land owners cannot keep cattle near to the border and the density of population is not sufficient to maintain a fly free barrier on its own without Government help," read one report.[16]

Apparently the Rhodesian authorities had a "fragmentary" record on this advance, until 1936 when new information showed a "serious threat" approaching the border twenty-five miles southwest of the known limits of the North Muzvirizvi fly belt that Swynnerton had written about in 1921.[17] From 1936 to 1948, the clearing was widened and improved, and in 1956–1957 it was extended to Beacon 99 along the Muzvirizvi River valley.

During the 1950s European farms came under increasing attack. There was a lull from 1948 until 1951 when fresh infections hit Mwangazi and Muumbe cattle dipping tanks, or "dips." The next year the first case was discovered at Muumbe (March) and again at Mwangazi (April), then two more cases for the rest of 1952. All told, forty-three cases were recorded at Muumbe, eighteen at Mwangazi in 1952–1953.[18] In 1953–1954, cases continued to be recorded at these dipping tanks, and the animals were kraaled and treated.[19] It got worse: new infections were now occurring further south on the Save in Ndanga East Reserve (March) and the Chisumbanje old dip (July).[20] The Muzvirizvi-Save River lowveld stretch along the border was deemed subject to invasion from Portuguese East Africa by *G. brevipalpis*, *G. pallidipes*, and *G. morsitans*. In April 1955, a new outbreak was recorded to the west in five cattle dipping centers. Two of them, Kondo and Chibungwe, served cattle that grazed on the escarpment, the rest (Gumira, Dakata, and Chibuwe) those that pastured on the Save. That very same year trypanosomiasis was diagnosed on Humani Ranch for the very first time. Five of these cases happened from March to May 1955.[21]

On March 24, Director of Tsetse Operations J. K. Chorley submitted a memorandum to the Trypanosomiasis Committee, detailing measures to deal with "what appeared to be a rapidly deteriorating trypanosomiasis position in the southern portion of the Chipinge district and the southeastern corner of the Ndanga district. Chorley requested permission to fence the international border from the Save River to Beacon 104; to extend the shooting area south from the Mkwasine-Save junction to the Musaswi River so as to cover the infected area of Mutandahwe Hill; to remove "all native stock" from the Save's east bank between Chisumbanje dip and Hippo Mine to the south; and to control all human and nonhuman traffic from the Mozambican border on the road to Mahenye. Members of the committee held the request in abeyance pending personal site visits.[22] Subsequently, a high-powered delegation including Chorley himself conducted a field visit. Entomologist Robert Mowbray was already there doing a tsetse and ecological survey of the area along

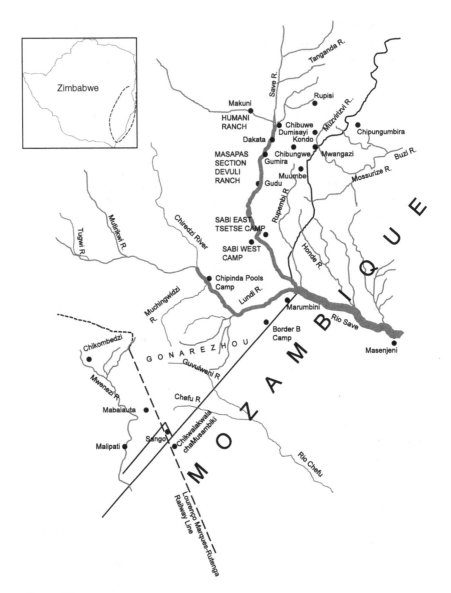

Figure 4.2

Tsetse advances into Chipinge and the Save-Lundi junction.

Source: Author

with the tsetse fly ranger in charge of Ndanga district. Both joined the inspection team.

At a subsequent meeting, a single three-strand fence was proposed "running over poor land" from near the Lundi-Chiredzi junction west to the Tugwi River, as well as a five-mile-wide "game-free cattle-free" buffer zone east of this fence. All Trypanosomiasis Committee members agreed except Chorley, for whom the proposal went against the successful precedent set in the Hurungwe and Nemakonde areas and would not halt the advance of the fly:

In my opinion it is not possible to create a game free-cattle free buffer zone using only a single fence composed of three strands of wire. Game within the area includes elephant, buffalo, eland, sable, roan etc., all of which would ignore the fence. . . . The whole scheme would break down owing to the necessity to provide an outlet for the native stock in the Ndanga East and Sangwe Reserves. . . . A five mile wide buffer zone is not wide enough to prevent the odd fly crossing it either on traffic or when naturally searching for food. The odd fly is known to range naturally further than five miles.[23]

Ultimately, it was Chorley's voice that his department superiors Gerald Cockbill and Desmond Lovemore listened to.[24]

It was at this time that Mowbray produced his reports, showing that mixed G. *morsitans* and G. *pallidipes* species had established themselves south of Mt. Makosa, while the upper Rupembe and its tributaries were "lightly infested" (Mavhunga and Spierenburg 2007). He determined the cause of the problem as one of human mobilities: species "are being carried on traffic from Portuguese East Africa in considerable numbers." Chorley elaborated on the mechanisms of human-tsetse transboundary movement, or the advance: "It is almost certain that these carried fly are responsible for the cases of trypanosomiasis occurring between the Hippo Mine and Chisumbanje. It is also possible that some of these are carried across the Sabi [Save] River on traffic. When the [Save] is low there is considerable traffic across the river at the P. & O. crossing and at the Sangwe Clinic crossing. Cattle from both sides of the river intermingle."[25]

In 1955, another survey was carried out, again on the Rhodesian side. This time one G. *pallidipes* was caught eight miles west of Save, and it was confirmed that G. *morsitans* was now established on the river. Another "burst of infection" was verified to exist north in the Muzvirizwi basin.[26] The situation escalated in April 1960 culminating in serious outbreaks among cattle pasturing along the escarpment that also dipped at Chibungwe and the new dip at Dumisayi. Meanwhile, on the west bank, the Masapas Section of Humani Ranch reported new cases that September. The confinement of infections to individual dips created the impression that two foci of infection existed, one on the escarpment, another along the Save River, either side of Chief Musikavanhu's area, which was cultivated and therefore formed a clear barrier to tsetse.[27] It was not known where the infection in Musikavanhu was coming from. However, it was thought to be "associated with traffic using the Mwangazi Gap through the hills." As Chorley admitted, this deduction was not based on any scientific experiment but the understandings and practices of local people: "Natives who have been questioned have stated that tsetse are occasionally seen on this road during the late dry season."[28]

The tsetse did not give the Tsetse Branch any respite in the 1960s. In February to April 1960, entomologist John Farrell surveyed the Save Valley, Tanganda River, and the Chipinga Plateau, along a line from Rupisi to Mt. Selinda.[29] South of Muumbe, in Lusongo area, one *G. morsitans* had been arrested in March, three in July, and another three in August. Trypanosomiasis outbreaks had occurred on the west bank of the Save that linked directly to the east bank owing to human mobility. In that area, the entomologist noted, "there is a large African population and considerable numbers of cattle" that dipped at Muumbe.[30] People refused to allow their cattle to be inspected at Mwangazi between July and October 1961 because that was always a prelude to inoculation with drug prophylactics that were killing them anyway.[31]

The search for tsetse's entry point(s) into Rhodesia deeply exercised the Tsetse Branch. Conventional wisdom was that the Mwangazi gap in the Save River was the main port of entry and officials readily envisaged "streams of tsetse and an unbroken chain of infection from [Mozambique] to the [Save] and points west."[32] Yet, as Chief Veterinary Officer William Boyt realized in 1964:

Nothing is farther from the truth. Between the western end of the gap and the riverbanks lies a belt of heavily settled and cultivated lands interspersed with mopani woodland. Cattle confined to this area were largely unaffected. . . . It seems more likely that the infection crossed to the South, to the river, in the region of Gumira and thence progressed along the eminently suitable riverine vegetation to Dakate and Chibuwe. . . . It may be that the infection ascended the west bank to Humani then infecting Chibuwe and Dakate herds which intermingle freely when the river is low with Mr. Whittal's stock on the west bank.[33]

This last point about the mobility of cattle in search of pasture and their vehicular role in the movement of tsetse fly and trypanosomes was for Boyt the critical factor:

A factor which may have a profound influence upon the ecology of trypanosomiasis in the area is the gross overstocking. The figures for the whole area stand at 28,000; the accepted figure is 12,000. . . . Such can only lead to bush encroachment and the creation of an environment more favourable to the tsetse fly. It also means the ranging of cattle further and further into the tsetse areas in search of grazing. This problem is largely of our making.[34]

Boyt was calling attention to mobility as risk and the need to establish a regime of discipline to tame it. But he was also suggesting that pasture was bad for tsetse control because it would force cattle to move beyond the cleared areas in search of greener pastures—and wander straight into tsetse-infested country. That was quite contrary to his predecessor John Ford's total belief in prophylactic settlement.

The Advance into Chitsa and Bikita

In 1941, cattle started dying on the upper Honde inside Rhodesia. That same year one *G. morsitans* was captured just one mile off the border.[35] The deaths continued into 1942; another *G. morsitans* was caught, this time seven miles north of the Save-Lundi confluence. Yet the six hundred cattle that watered there displayed no sign of infection whatsoever.[36]

As entomologists found in 1942, it was not just *G. morsitans* responsible for this; *G. pallidipes* also occurred along the border north of Chipung-umbira (called Espungabera by *maputukezi*) while *G. morsitans* pushed southwest along the Buzi. By 1942 the latter species could be caught on the Rhodesian border and had infected the Honde River valley.[37] In 1943, more cattle—160 of them—died. This time several tsetse flies were caught.[38] In the next three years the insect "advanced" rapidly.[39] West of Save, the first case of trypanosomiasis was reported in 1944, two years after the east bank outbreaks.[40] At the Save-Lundi junction, Chief Chitsa and his subjects reported four cases of trypanosomiasis among his cattle the same year. The situation "remained fairly static."[41] Then in 1945, "extensive" infections were witnessed on the upper Muzvirizvi forty-five miles north of the Save-Lundi foci. The strain was associated with the tsetse fly colony long anchored on the Buzi River basin of Muzvirizvi. South of the Rio Save, meanwhile, the tsetse had pushed to within just twenty miles of the Rho-desian border, a situation made more difficult by a foot and mouth disease (FMD) outbreak.[42]

In 1946 the cattle evacuations began: all surviving livestock on the Rhodesian side east of Save were inoculated before being driven north, out of the fly-infested area.[43] Donkeys were not included in the evacuation; later they were in fact deployed as sentinels to detect the presence of tsetse in the Murongwezi, Mabee, Makoho, Chisuma, Hippo Mine, Mwangazi, Mareya, and Zamchiya areas.[44] No losses of cattle were reported in the area until 1947, afterward followed by another four- to five-year respite.

The trypanosomiasis infections increased in the 1950s. In 1951–1952, a severe outbreak among Chitsa's cattle confirmed the authorities' worst fears—the fly was now well established west of Save River, its pre-rinderpest precinct. Chitsa's cattle were removed northward to Ndanga East Reserve.[45] In November 1952, Ndari dip reported its first cases, followed by two more in April 1953. In March 1954, cattle smears at Rupangwani (July) and Ndari (September) tested positive for trypanosomiasis. The situation had improved by January 1955 at Ndari, but definitely not at Muteyo, where nine cases were diagnosed that month. In April, further infections occurred at Rupangwani and Ndari, while Sangwe Reserve recorded its first case. At Gudo and Humani Ranch one and four cases were recorded respectively.[46] Throughout 1956, further cases occurred at Rupangwani, Muteyo, Ndari, and Humani. During 1957 to 1958, a clearing in the Bandai area had been completed, and it significantly reduced infections to "sporadic and seasonal . . .distribution" until 1963, when they shot up again from nine in 1961 to 29 in 1962 at Rupangwani, and seven in 1961 to 71 in 1963 at Peri.[47]

The Lundi Advance

It is inevitable that mobility—specifically the direction, pace, or actual visualizable movement—would be the site to look for the advance of the tsetse fly as seen and felt by the Rhodesian settlers. For one can only understand advance as a narrative of unfolding and enfolding events.

In the wake of the rinderpest, the white hunters discussed in chapter 3 often encountered the tsetse fly along the Rio Save inside Mozambique, in the same areas where forest animals like buffalo and elephant were most abundant. The big tuskers promised instant riches in ivory, but the problem was *nzira* (the way) and means to get the ivory out of the forest. Donkeys, while considerably more resistant to tsetse fly attacks than horses, could never make it out alive. Yet locals in the area still kept dogs, and these animals only succumbed during the searing droughts that plagued the area, not to biting tsetse. The local inhabitants kept these dogs alive through the "practice of feeding a few dead tsetse flies to all their animals" (Bulpin 1954, 58). White hunters, meanwhile, usually smeared a cocktail of poisons (arsenic), attractants (honey), and sticky stuff (tree gum whose harvesting and making was steeped in local practices). The donkeys were converted into "pretty sticky-walking fly traps" but they indeed survived (ibid.).

Initially there was no cause for panic. In 1918, Swynnerton had found the southernmost limit of the fly belt to be 20° 20′ (33° 25′ W), on the Buzi river area of Mozambique to the west of the Sitatonga Hills. In 1923, the chief veterinary surgeon of Southern Rhodesia reported cattle mortalities in Ndanga East Reserve and suspected it was tsetse fly (*G. morsitans*). Nothing was made of the matter until 1928, when the chief entomologist referred to the fly's presence on the Rio Save area of Masenjeni, just thirty miles into Mozambique. It was clear that the fly was pushing toward the Save-Lundi Junction, spread by the cross-border grazing of local maTshangana's cattle. In 1932, the chief entomologist warned that the proposed Gonarezhou game reserve was too near the border for "protection from the advance of fly." A permanent game reserve would be "a very short-sighted policy" because thousands of ranch cattle that grazed annually on the "buffalo grass" north of Chipinda Pools along the Chiredzi River were far more important than forest animals (Mavhunga and Spierenburg 2009).[48]

Between 1925 and 1929 Rhodesia had persistently asked the Portuguese to verify the tsetse fly position and take protective measures, but the *maputukezi* were preoccupied with the strain in Tete and Cabo Delgado districts (Kruger National Park 1933, 64). Furthermore, laboratory samples from cattle with "trypanosomiasis-like" symptoms and insect specimens captured in Gonarezhou tested negative.[49] The Rhodesian authorities had a "fragmentary" record on this advance, but by 1936 they viewed tsetse as a "serious threat" to Save valley. As Swynnerton also found, *G. pallidipes* occurred along the border north of Chipungumbira while *G. morsitans* pushed southwest along the Buzi. By 1940 tsetse fly had effectively arrived in the Save-Lundi area.[50] By 1942 *G. morsitans* could be caught on the Rhodesian border. In the next three years it advanced rapidly: in 1943, two flies were caught at Mahenye on the Lower Rupembe, and by 1945, the herds of Chief Mahenye and his maTshangana people living along the Mozambique-Rhodesia border had been decimated from six hundred to just ninety-four head.

By 1944, tsetse fly was well established in its pre-rinderpest haunts between the Save and Lundi, and Nuanetsi Ranch and the Transvaal now faced a real veterinary threat. In 1944, a single fly was caught west of the Save; Chief Chitsa, who lived with his subjects in the Save-Lundi, reported four cases of trypanosomiasis among his cattle the same year.[51] Fencing the boundaries was now a matter of life and death for the white settler cattle industry. By the end of September 1945, anti-tsetse shooting reached the Gonarezhou reserve and was heading with speed toward the Mwenezi River. The tsetse fly and trypanosomiasis situation remained unchanged until 1947, when foot traversal along the south bank of the Rio Save encountered neither tsetse fly nor sick cattle as far as twenty-five miles into Mozambique. There was a palpable sense of tense relief in Salisbury. Apparently the Portuguese had quarantined all cattle south of the Rio Save, a few miles east of the international border. The animals showed no sign of trypanosomiasis, confirmation that the south bank was fly-free except for the fifteen-mile radius of Masenjeni.[52]

In 1949, an entomologist named H. E. Hornby was again dispatched to undertake a thorough survey of adjoining Portuguese territory east of the Save-Lundi junction while two tsetse field officers made "cursory examinations" of the Rhodesian side.[53] In July 1950, a game ranger named Hooper caught fifteen *G. morsitans* "somewhere west of the border"; exactly a year later (again in July) *G. morsitans* were reported at the Shabani Mine recruiting station at Marumbini.

In 1952 *G. pallidipes* was caught a few miles from the border, as the government drove all cattle out of the Save-Lundi junction to a safe distance further north.[54] The plan was to create "a deep defensive zone."[55] In 1953 the first *G. pallidipes* on the Rhodesian side was taken east of the Save at Chivirira Falls.[56] Meanwhile, "heavy mortality" in cattle was experienced in adjacent areas of Mozambique; in fact by March 1953, all cattle watering on the Rio Save had vanished, with the only herds now left at Kapitani's.[57]

Local African inhabitants felt the tsetse's presence through the deaths of their cattle. Says Jerry Masevhe: "This tsetse . . . entered through Chileji (Chiredzi River), eh, in the bush they began to see cattle just dying, dying cattle. They said 'Hey, there are marauding flies terrorizing our cattle, they are being carried from Mozambique by the wind.' After a while another was caught."[58] The "wind" was of course people coming from Mozambique to Rhodesia to look for work in the farms, mines, and emerging towns.

The situation was deteriorating; *G. pallidipes* had settled down on the Honde River. In April, Chorley gave instructions to entomologist K. E. W. Boyd to go and "put a finger on the pulse of the fly" on the Save's west bank between the Mkwasine and Lundi river junctions.[59] Boyd was one of several *valungu* to arrive in the Save-Lundi area and enlist local vaShona and maTshangana men to catch tsetse so that he could examine them.[60] They enlisted as flycatchers to earn money to pay taxes and eke out a livelihood rather than to advance the cause of science.[61]

Boyd found no fly, so another survey was ordered in August. This time the entomologist apprehended the little fellow on the Save's west bank close to the junction with the Lundi, the site of cattle evacuation in 1952.[62] Clearly a "major advance" of tsetse had occurred southwest of Masenjeni. In his 1954–1955 report, the Director of Tsetse Fly Operations disclosed that the Chivonja plateau was "probably permanently infested with tsetse."[63] In 1956 G. pallidipes and G. morsitans continued their advance west, covering ten more miles. Twelve months later they were still there. In 1956, a survey yielded just one fly; the following year, one more.[64] Ford concluded: "the advance thus pushing out in various directions represents a movement towards restoration of the position prior to the Great Rinderpest of 1896."[65]

It got worse. In May 1958, G. morsitans had occupied the areas either side of the border road to thirty miles northeast of Malvernia (Sango) along the Lourenço Marques (Maputo) railway line and just sixty miles from the Limpopo River.[66] In July, cattle watering on the Chiredzi River tested positive to trypanosomiasis; this as G. austeni was identified as part of the traveling party of tsetse flies from Mozambique. Tsetse fly was also apprehended on Portuguese African workers on the Save River west of Masenjeni, while one was taken at a de-flying chamber right on the border "on a traveller moving from Rhodesia into Mozambique" despite the G. austeni epicenter lying deep inside Portuguese territory.[67] That same month came the alarming finding that tsetse was on the move northward from Chivonja Hills toward Lone Star Ranch, located east of the Chiredzi River. Also that July, two positive cases of trypanosomiasis were found among eighty-nine smears taken from a herd that had been grazing in the northeastern corner of the ranch for eight months. In November, another case was discovered among cattle watering on the Lundi, while four more positive cases were diagnosed among African cattle in the Chikomba area. The cattle watered on the Chiredzi between the Nyamasikana and Chikokovela Rivers, while the European cattle grazed in the northeastern parts of the ranch, at the Caves kraal, Artesia, and Chipanguchi Pan, on the Fair Range boundary, and southeast near the Lundi River.[68]

The first infection appeared southwest of the Lundi in July 1958 at Chibwedziva dip. Fifteen cases of severe trypanosomiasis and rapid mortality occurred; afterward the severe cases "relapsed into the more usual chronic manifestations of trypanosomiasis vivax."[69] The infection was restricted to herds grazing and watering in the vicinity of the Muchingwidzi and Lundi Rivers.[70]

The situation rapidly deteriorated after 1959, with no letup despite bush clearing and the construction of a new dip at Chambuta in 1962. In fact, in 1963, Chilonga dip (north) and Machindu (central) and the area near the road to Chikombedzi became infected. Chilonga dip was on the Lundi and marked the northern limit of the trypanosomiasis strain up until then. In the same year, the Matibi II Reserve was extended southward to include areas served by dips at Matigwa, Mpapa, and Masivamele, all later confirmed infected. The three thousand head of cattle people brought in were now exposed to infection; the trypanosome had found a new vehicle. Meanwhile, a South African team conducting a pupal survey confirmed that tsetse fly had moved as far as the Gubulweni River and there was now "a very real threat to the [Mwenezi] River and the extensive ranching country beyond."[71] It prompted the setting up of four test (or sentinel) herds of cattle, two near the Malvernia railway line near Tswiza and Nyala sidings, and two on the middle and upper Gubulweni. After just a fortnight, infection was recorded at Tswiza, but not at Nyala. The rapid spread of trypanosomiasis spelled disaster for Nuanetsi Ranch just further west. By this time, Gonarezhou had been proclaimed a game reserve amid stiff objections from entomologists.[72]

The southward advance seemed to fizzle out in 1959 but not the westbound one. Now most of the Nyamasikana valley was infested with tsetse, with two *G. morsitans* caught on a motor vehicle at a checkpoint set up along the railway thirty miles northwest of Malvernia.[73] Between July 1958 and March 1959, no cases were found among European cattle; in any case, any that showed clinical signs of trypanosomiasis were inoculated. It was among these cattle that nine cases of infection were diagnosed in March 1959, rising to fourteen the next month. Of one thing the Tsetse Branch was certain: "The continuing infection suggested that the riverine woodland of the Chiredzi was harbouring tsetse."[74]

Conclusion

The tsetse invasion demonstrates one clear fact: that European and human colonization of the borderlands was never quite complete—if such colonization ever happened at all. Over and above illustrating Kjekshus's argument that Europeans disrupted ecological regimes that Africans had established, and that those regimes had been crucial in containing the mobilities of disease-causing pests, I have sought to draw attention to the mobilities of the insects themselves. This is just the beginning of a conversation carried through into chapter 5. For the question that now confronted the state was: How then to control, contain, or eliminate the mobilities of the tsetse fly?

What the European settlers called the "advance" of the tsetse fly was a sign of the animal population's recovery and expansion into habitats beyond the colonial margins. This is an example of an insect's role as a self-propelled vehicle for the trypanosome and a passenger catching a ride on anything that moves, on mobile animal and domestic animals (the fly's food source), and on people going about their business on foot, bicycles, and motor vehicles. The insect, the human, forest animal, and the car all become self-propelled or auto-mobile vehicles transporting the tsetse fly and enabling it to invade. This chapter has sought to appreciate the productive power of organic mobilities that, in the next chapter, had to be stopped dead in their tracks through gunfire from the very same people the colonial state now criminalized as poachers: the African professoriate of the hunt.

5 The Professoriate of the Hunt and the Tsetse Fly

Until the appearance of trypanosomiasis drugs and residual insecticides in the late 1950s, there was no other method besides "game destruction" to starve the tsetse fly and deny it its primary vehicle for mobility: big forest animals.[1] Similarly, in the absence of ready-made technological solutions to tropical pests and pestilence, Rhodesian settlers turned to African huntsmen and deployed them as a weapon of tsetse control. The term "game" becomes appropriate to use in the reference to how white settlers viewed these animals.

The prospect of walking through the "bush" in rugged terrain at the mercy of all kinds of insects, snakes, and predators, in punishing heat in remote areas with no other white soul in a radius of fifty miles, was not likely to attract too many *valungu* to anti-tsetse operations. At most, a few candidates were found that were interested in a job as tsetse ranger, but few could stay even for a couple of months—if the fever spared them that long to start with. This is how European colonizers who came to Africa, convinced that they were engaged in a civilizing mission and using science as a tool, discovered that their tools did not work and so they came to depend more and more on what Africans knew and used (Tilley 2011, 195).

In this particular case, the state turned to the *hombarume* or *maphisa* who lived nearest the tsetse-infested areas. It armed them with Martini-Henry rifles and .303s and sent them into the forest, and made it a policy that the more animals they slaughtered the more they would be rewarded. They inserted their "individual temporalities and local rationalities" (Diouf 2000, 679–680) into what was supposed to be the colonizer's veterinary and entomological project. The government-issued guns and ammunition became instruments for the provision of a dietary need—*nyama* [meat]—now that Africans were banned from hunting or possessing guns.

Both the gun and the person using it became technologies of tsetse control to the state, as the state and its guns became means of acquiring meat and continuing the hunt to Africans. We have already seen this story play out in the *migodhi* (mines), where African aspirations for self-imagined and self-made modernities, and the imperatives of mining capital were simultaneously met through the sweat of the African migrant worker. The *work* of securing settler cattle wealth was performed through the mobilization of Africans as "human motors" (Rabinbach 1990) equipped with tools (guns) to eliminate the vehicles (animals) that carried the trypanosome and fed the tsetse fly. What Aimé Césaire called thingification—the colonizer's transformation of Africans into things, in this case machines (Césaire 1955/2000, 42–43), encountered an African who thingifies back. The transient work Africans did with guns for the state and for themselves gives "a radically different picture of . . . invention and innovation" and "a whole invisible world of technologies" (Edgerton 2007, xii), forcing us to rethink what constitutes technology and use when the state and its subject use each other as means and ways to ends.

As a problem to be solved, the tsetse and the trypanosome it carried provided what MacNaughten and Urry (2000, 169) call an *affordance*, a space within which hunters became technology and hunting itself a cleansing kind of movement. This hunter tracking through the forest on foot with gun in hand, firing and producing carcasses that meant different things (tsetse vector elimination to the state, meat to the hunter) is another iteration of the transient workspace. The forest animals, wherever they may have been, were the transient workspaces of the tsetse fly, the site where it was following, buzzing around, dipping its proboscis into the animal's body, and drawing and ingesting blood contaminated with trypanosomes that it carried around and deposited in livestock. This is how the site of big animals became the workplace where the hunter performed his work, the primary objective of which was to deny the tsetse fly its capacity to be mobile. A forest animal became vermin not merely by the act of a tsetse fly feeding on its trypanosome-contaminated blood. Rather it was the fly's ingestion and transportation of this protozoan and its inoculation into livestock that made forest animals vermin and the tsetse itself a pest.

Rhodesia's Anti-Tsetse War: A Brief Overview

In 1919, barely a year after Charles Swynnerton's Mossurize study, "game elimination" began on an experimental basis in the Gwaai and Shangani Native Reserves of Sebungwe District with the aim of finding "a practicable method of fighting this terrible scourge" of tsetse fly.[2] By 1920, the shooting experiment was exhibiting positive effects; "a considerable reduction of fly," especially in areas where animals had been "most effectively reduced," was recorded.[3] On December 15, 1921, the Gwaai-Shangani experiment was wound up, with the chief entomologist declaring that the shooting had resulted in "a very marked reduction of tsetse." That same year plans were drawn up to carry out similar experiments in the Mupfure River area of Chegutu.[4]

In 1923, game elimination was extended to the rest of the country as the first line of defense against tsetse fly invasion. Following deaths in numerous "native cattle," shooting operations were extended to the Kandeya area of Rushinga and Nemakonde in November.[5]

The policy of game destruction generally continued unchanged until 1956 when all shooting was restricted to specific species and in less threatened areas was discontinued altogether.[6] The process of replacing game destruction with other techniques gathered momentum in 1959 and 1960. All guns fell silent in the Save River valley in January 1960, giving way increasingly to organochlorine pesticides like dieldrin, DDT, and endosulfan.[7] Game slaughter otherwise continued unabated in Sebungwe, barely dented the fly density, and lent even more weight to advocates of pesticides.[8] In 1960, the Director of Tsetse Operations conceded that the two-and-a-half year onslaught had "made no impression on the tsetse population." Worse, the smaller animals not only persisted, they actually increased.[9] In Hurungwe, game eradication was also terminated even when the director thought it would work if applied correctly.[10]

1961 was a decisive year for three developments. The first was a moratorium placed on game elimination as a method of tsetse control, as fears of game extinction grew. DDT and other pesticides targeting the tsetse fly personally would be used in conjunction with erecting tsetse control fences. Second, all "untrained hunters" (that is, those recruited on the basis of knowledge and skills acquired in the professoriate of the hunt), were discharged and replaced by Department of Tsetse and Trypanosomiasis Control-trained "field assistants" conducting bicycle patrols and flyrounds.[11] Third, having operated as an independent government department, tsetse control was folded into a branch of the Department of Veterinary Services within the Ministry of Agriculture. Another department—the Department of National Parks and Wild Life Management (DNPWLM)—was also created to oversee the conservation of forest animals in national parks.[12]

The moratorium on game elimination backfired badly. From 1961 to 1964, animals dramatically recovered and reoccupied cleared areas, and with it came tsetse fly reinvasion. Most of the gains made up until 1960 were erased. In 1964, the state reversed course and authorized the use of "all available methods in future, including selective game elimination."[13] The tsetse fly areas were divided into eight "controlled hunting areas" (CHA): North and East Hurungwe, Nyaodza, Save-Lundi, Sebungwe, West Hurungwe, The Gokwe-Sanyati, North Sipolilo, and North Nemakunde. DNPWLM rangers and selected Tsetse Branch officers were directed to remove all elephant and buffalo from these CHAs. All favored hosts of tsetse (warthog, kudu, bushbuck, and bushpig) were to be shot and elephant and buffalo driven away from the tsetse control fences.[14]

All these operations continued in the various districts from 1966 to 1970 with positive results. Gone was the earlier method whereby hunters were stationed in teams of twos and threes, at fixed camps, distributed over wide areas, from which they operated for each hunting period. In their place was a "mobile team" composed of one European tsetse field officer, one African supervisor hunter, and twenty-five African hunters. They were issued with rifles and ammunition at dawn, to be surrendered at dusk, with all ammunition being counted before and after each workday. The new technique was hailed as "an advance" for the "very much greater control of the hunters, thus making it possible to manipulate the hunting pressure" on tsetse to a greater extent than hitherto possible. Indeed, Tsetse Branch boss Gerald Cockbill could declare in 1971 that "the normal [tsetse] pattern had been upset . . . and that the disease position was improving."[15] The onslaught continued.

The Southeastern Front

In the southeastern corner of Southern Rhodesia, the "tsetse invasions" took the shape of a westward movement from contiguous parts of Mozambique. These insect mobilities were not simply the result of forest animal populations recovering from rinderpest and returning to their old haunts. Rather, this refuge mobility carrying tsetse along with it was an outcome of the opening up of the *zonas de caça* or *coutadas* (hunting areas) adjacent to the Southern Rhodesian border in the 1930s.[16]

Inevitably, from 1934 onward these animals served as organic vehicles purveying pathogens into Southern Rhodesia. A corresponding series of East Coast Fever (*theileria parva*) and foot and mouth disease (FMD) forced the cancellation of plans to declare a game reserve around Chipinda Pools and to annex the whole Gonarezhou forest into a vast transfrontier game reserve joining up with Kruger National Park and adjacent *zonas de caça* in

Portuguese East Africa (Mavhunga and Spierenburg 2009). In 1945, hunters on tsetse control duty were withdrawn to facilitate veterinary measures.

The animals also purveyed the tsetse "invasion" from the Mozambican precincts of the Muzvirizvi and Rio Save areas of Masenjeni, the little insect riding on the backs of elephant and buffalo fleeing trophy hunters and *Chefe de Postos* into the refuge of Gonarezhou. The revival of big animal populations like buffalo and elephant in the area between Chipungumbira and Chibwedziva marked the beginnings of trypanosomiasis outbreaks (Mavhunga and Spierenburg 2007).

South of Rio Save toward Crooks Corner and Kruger National Park, buffalo fleeing the shooting was feared to carry FMD, not tsetse fly or trypanosomes.[17] In fact, so insignificant was tsetse fly south of the Lundi River that in January 1956, tsetse control-related hunting was discontinued in Chiredzi, with all resources being redirected to cutting a fence line and fence maintenance roads.[18] Instead, the game elimination operations were confined to the north of the Save River in the Ndanga and Chipinge areas. The hunting operations in the fenced area south of Ndanga East started on April 1, 1957.

The area had been partitioned into six sections, each serviced by a hunting camp for the European tsetse ranger and readily accessible by road to facilitate his control of African hunters. One "supervisor hunter, 12 hunters and two labourers" were stationed at each camp. The "supervisor hunter" in this case was a European, the rest Africans. The twelve African hunters operated in pairs since they came from the same village and had previously hunted together in the professoriate prior to recruitment for tsetse operations.[19]

In 1957, it was resolved to focus the whole "hunter force" on the Mt. Makosa area, which had long been "the major source of animals on the east bank." Since 1940, hunters had slaughtered many an animal in the area, but they were too few, the animal population remained high, and so did the density of tsetse. Getting enough good African hunters was difficult, and the force was often down to half strength. To conquer the fly, 1,800 animals had to be eliminated every nine-month shooting season. One season into the program the Department of Tsetse and Trypanosomiasis Control had a 200-head shortfall.[20]

In June 1958 forty hunters were concentrated in the eighty square-mile area of the hills, with five hunters armed with shotguns joining them from November. The effect was instantly felt. In January 1960, the program was shut down throughout the southeastern tsetse area in line with national policy. The tsetse recovered. When operations resumed in 1964, the invasion had spread south of the Lundi.[21]

Then, in 1965, the highest echelons of the African nationalist leadership were incarcerated at the Gonakudzingwa Detention Center in northern Gonarezhou (see figure 7.4), and their subversive activities set on fire African emotions in the areas of Sengwe, Maranda, Ndanga, and Chipinge. In 1965, the same year that Rhodesian white settlers issued their unilateral declaration of independence (UDI) instead of handing over power to the African majority, the government declared a state of emergency in the districts around Gonakudzingwa. Since Africans were in revolt, it was not wise to issue rifles that might be used for insurrection. Security authorities therefore laid down conditions regarding the strict supervision of hunting teams; fifteen additional Tsetse field officer posts were created for the purpose in 1966–1967 alone.[22]

Progress from 1967 to 1968 was pleasing. In the face of a combined gun and insecticide onslaught, *G. morsitans* vanished, barring a few stubborn colonies.[23] In the northeast-southwest barrier extension east of the Save River, shooting operations were terminated in October owing to extreme animal scarcity. The killing fields of the Save-Lundi junction yielded reduced carcasses from the previous year's numbers, testimony to the low populations of selected animal species still standing. At this point the Tsetse Branch decided the time had arrived to reduce the numbers of hunting teams from three to two for 1968.[24]

The 1968–1969 operations in the Save-Lundi involved three teams until January 1969 when one team was withdrawn following "further substantial reduction" in animals killed—from 455 the previous year to just 270.[25] Only two teams were still shooting in the Gubulweni-Chefu area and the northeast-to-southwest barrier extension. The numbers had shrunk even further to just 197.[26] The last holdouts of the unwanted animals were diminishing.

What Type of Guns Were Africans Issued?

By 1934 significant technical developments were forcing the state's hand to switch from the Martini-Henry to better, more up-to-date models of guns.[27] First was the shortage of (cartridge) ejectors, and the African hunters were asked to improvise, meaning, use their ramrods to remove them instead. (The extractor is what pulls the cartridge out of the chamber after firing, and then the ejector kicks it out of the firearm.) Increasingly the Martini-Henry became unserviceable,[28] its ammunition scarce and obtainable only from leftover government stocks in Europe.[29] In Rushinga, the number of African hunters was cut by half in 1937 since operations only involved consolidation.[30] The Department of Defense had ordered ammunition from the United Kingdom in 1938, but hostilities broke out in Europe the following year.

Ammunition was a problem not least because some African hunters, who had hitherto used muskets, were still getting used to the Martini-Henry rifle.[31] Authorities proposed two ammunition-serving options: to either reduce the number of hunters, or else restrict the number of rounds issued each month. The latter option prevailed—a 25 percent cut in monthly allocation was imposed.[32] Chief Entomologist J. K. Chorley acknowledged in his annual report of 1938 that "the reserve supply of Martini ammunition became dangerously low and, in consequence, the number of native hunters had to be reduced considerably."[33] Later that year new supplies were received, and operations were intensified again.[34]

The problem of ammunition continued to hog the limelight during the 1940s. In 1940, a total of 15,657 animals were destroyed for an expenditure of 34,447 rounds of ammunition or 2.2 rounds per head.[35] In 1941, a total of 20,512 head were destroyed with 43,286 rounds of ammunition, or 2.1 rounds per head.[36] In 1944, the kill rate (number of bullets expended per kill) was 2.2 rounds per head or 27,272 animals destroyed for an expenditure of 59,195 rounds of ammunition.[37] In 1945, 28,086 animals were killed for the expenditure of 58,843 rounds of ammunition, or 2.1 round per head.[38]

Chorley favored a switch from breech- to magazine-loading guns. The Winchester .35 rifle was his proposition if something larger cost more. The Prisons Department already had those. But no stocks either of the rifle or its ammunition were available in the country. It took nine months to ship from the United States in peacetime. The real reason for disqualifying the Winchester, however, had more to do with the rifle's technical properties: it broke easily at the stock, required constant oiling and attention, and the magazine was easily dented, thereby completely disabling loading.[39] The chief entomologist was worried that if these hunters were issued with a magazine rifle "with only moderate hitting power, the consumption of ammunition may be doubled or trebled without any increase in the number of game destroyed." He would rather take "any surplus of old single loading .303 rifles" from the UK's War Office or else approach countries with excess Martini-Henry ammo.[40] He finally got his first option but almost two decades later.

The decision to switch from Martini-Henry to .303 rifles was taken because stocks were running low and a number of rifles were wearing out. The .303 ammunition was "readily available while Martini-Henry would have to have been specifically manufactured at much greater expense."[41] In November 1946, as a result of the shortage of Martini-Henry ammunition, the number of rounds expended each month was cut in half. The destruction of the smaller animals, including steinbuck, Sharpe's steinbuck, oribi, klipspringer, blue duiker, and Livingstone's suni was stopped. 24,351 head had already been shot with 53,576 rounds spent, or 2.2 rounds per head.[42] The shortage of Martini-Henry .45 ammunition forced the tsetse control department to reduce the number of rounds of ammunition expended each month by 50 percent beginning in November. The destruction of smaller animals such as steinbuck, Sharpe's steinbuck, oribi, klipspringer, blue duiker and Livingstone's suni was suspended.[43]

In 1947, the number of animals destroyed was unsurprisingly reduced, "partly due to a temporary shortage of Martini-Henry .45 ammunition, and partly to the decision that the time had arrived when smaller animals could safely be spared." 16,802 animals were destroyed with 32,996 rounds of ammunition, or 1.96 rounds per head.[44] The number of animals destroyed was also considerably reduced owing to the ammunition shortages and sparing of smaller animals.[45] Resulting from the combination of ammunition shortages and exemption of smaller animals from elimination, the number of animals destroyed dropped to 16,802 head for 32,996 rounds of ammunition, or 1.99 rounds per head.[46]

In 1948, with the restoration of supplies of Martini-Henry ammunition, the number of animals killed increased. But the average rounds per head had increased: 2.25 rounds of ammunition per head (for a total of 22,160) destroyed.[47]

In 1949, the kill increased to 24,871 due partly to the intense shooting in Sebungwe and partly to the reintroduction of shooting duiker in that area. 2.29 rounds of ammunition were expended per head.[48]

Normal supplies of Martini-Henry .45 ammunition resumed in 1947. The exemption of smaller animals was continued since it had had no significant effect on tsetse eradication. There was a slight increase in the number of animals shot—22,160 head compared to the previous year's 16,802, but still less than the 24,351 shot in 1946. This was an average of 2.25 rounds of ammunition per head destroyed, or 55,400.[49] The following year 24,871 head were killed for an average 2.29 rounds, or 56,954 rounds total. The increase was due in part to the intensified shooting in Sebungwe and in part to the reinstatement of duiker on the hit list.[50]

The 1950s were marked by the first changeover in the guns—and inevitably ammunition and shooting skills—from Martini-Henry to .303 rifles. In 1952, the tsetse control department retired Martini-Henry rifles and armed the African hunters with Lee-Enfield .303 rifles. Because their accuracy with the new rifles was less than with Martini-Henrys in this transitional period, Africans were forbidden from shooting elephant "because of the danger of leaving them wounded."[51]

The changeover from Martini-Henry rifles to the Mark VII .303 rifle was completed in the 1952–1953 operational year. The rifle's spare parts were difficult to obtain, so it was resolved to switch to the old Mark VI carbine readily obtainable within Southern Rhodesia. This was only a temporary relief given that "spare parts for this type of rifle are nonexistent."[52] In the 1952–1953 operations, 51,216 rounds of ammunition were expended on the slaughter of 27,535 animals (or 1.86 rounds per head). The figures excluded the numbers of animals that European hunting parties killed while operating under permit.[53]

In 1953, by arrangement with the Department of Defense, 660 Mark VII .303 rifles were withdrawn and African hunters reissued with Mark VI .303 carbines. Only three operational areas continued to use the Mark VII.[54]

In the 1953–1954 season, 95,999 rounds were expended in the destruction of 36,910 head, or 2.6 rounds per head. This 9,529 head increase was a result of the general intensification of operations, individual efficiency, and more effective control of the African hunters.[55]

In 1954–1955, a total of 800 hunters were deployed throughout Southern Rhodesia's tsetse-infested areas. They killed 41,076 head, using in the process 107,401 rounds of ammunition or an average of 2.5 rounds per head, 4,666 more than the previous year. The total number of animals eliminated in the 1955–1956 operations was 34,208 animals for an expenditure of 101,275 rounds of ammunition, or 2.96 rounds per animal. The higher round per animal ratio was attributed to the small game, constituting 64.6 percent of total kills, with elephant restricted to European officers, and not to the hunters themselves.[56]

From 1958 onward, game elimination entered a new phase: the introduction of shotguns alongside the .303s. The switch was designed to solve the problem of eliminating smaller game species like reedbuck, duiker, steenbok and Sharpe's grysbok. The hunting operations were carried out on moonless nights using three-cell torches mounted on the shotguns. Each African hunter was assigned an African "laborer assistant" both to carry supplies and "for morale" in the scary, snake-infested forests. Unlike their .303-armed counterparts who operated from "fixed camps," shotgun hunters were moved frequently around to areas of small animal sightings.

To get a sense of how effective shotguns were measured against .303s, an average 9.11 head per hunter per month fell to the shotgun compared to 3.99 head per .303-armed hunter per month in 1958–1959. The shotgun's vast superiority was most pronounced in the dry winter months, particularly after all grass was burnt, the target clear. In September 1959 for example, shotgun hunters killed 15.08 animals per hunter compared to just 4.82 per regular hunter with .303s for all animal types. The difference for smaller animals alone was astronomical. In the same period (1958–1959), the hunters used just 1.69 rounds per animal to kill 765 animals, while regular .303 hunters expended 3.99 rounds per animal when killing 6,045 animals.

The shotgun must be seen as a complement rather than replacement for the .303. It was effective only for small animals and just about worthless for anything else. Hence it always had to be used in conjunction with rifles, not as replacement for them. The numbers of *makudo* (baboon), *mhumhi* (wild dog), *mbizi* (zebra), *nguruve* (bushpig), *njiri* (warthog), *nondo* (tsessebe), *mhimha* (reedbuck), *nhoro* (kudu), *mhembwe* (bushbuck), and *mhara* (impala) killed using the shotgun paled into insignificance compared to its devastating effect against duiker, steenbok, and Sharpe's grysbok.

In November 1958, the Department of Tsetse Fly Operations authorized night hunting with shotguns and spot lights in Sebungwe. It soon became apparent that a bright light frightened away such animals like the reedbuck, so the hunters used a weak light instead, or were authorized to poison "all isolated surface water with diesoline and DI-IC (surplus from the aerial smoking campaign of 1957)." That way, all water would be rendered unpalatable and the animals would starve to death and be forced to concentrate at pools where spotlights could be used to aid shotgun-armed hunters.[57] The positive results convinced tsetse officials that "the regular hunters had not been effective in curbing these populations" and that the time had come to hand out shotguns to "the best regular hunters," arrange them into squads and deploy them to cover distant areas of limited hunting intensity. The operation began at last light and continued to around eleven. The results showed that, based on the number of animals shot per hunter and the shots per kill, shotgun hunters were "more efficient." But it was not that simple:

Although the all-round efficiency of hunting with shotguns was higher . . . each shotgun hunter shot an average of only 4.25 head of the tsetse's primary food

animals, whereas his regular counterpart shot 11.5 head in the 12 month period.
. . . Shotgun hunting was particularly effective against small antelope in more open
habitats as the hunters were reluctant to enter thickets at night. Besides [bushpig,
warthog, kudu, and bushbuck] these included reedbuck and perhaps impala. All
adapted quickly to the new form of hunting and within a few months were difficult
to locate with a spot light from a vehicle before the hunters went to bed, although
afterwards they could be found in numbers.[58]

In the 1959–1960 period, the number of regular .303 hunters varied
from 111 to 123 compared with 120 to 130 in 1958–1959, while shotgun
hunters increased from 12 to 23 in March and then to 50 in April compared
to the increase of 5 to 12 in 1958–1959. The shotgun increase improved
the ability of hunters to cover the ground thoroughly, especially in the
Cewali [chiware] section, where twenty-five of the shotgun hunters operated. These hunters also brought full weight to bear on the smaller animals.
The African hunters destroyed 10,033 animals, the shotgun hunters 3,951,
or 1,039 animals per hunter per month, while .303 hunters shot 6,082
animals or 3.94 animals per month. Shotgun hunters expended 1.85 rounds
per animal, while those with .303s shot 3.12 rounds per animal. The
increased haul of carcasses was a result of intensified shotgun hunting by
night, although .303 hunters accounted for 3 percent more head of small
animals than they did in 1958–1959.[59]

Shotguns were ideal for attacks on the smaller animals. Ammunition
expenditure was 1.85 rounds per animal for shotguns and 3.12 rounds per
animal for .303s. Despite the .303 hunters accounting for 3 percent more
head of small animals than they did in 1958–1959, it was obvious that
hunters had a better kill rate with shotguns than with .303s.[60] In Sanyati,
the average was 7.2 animals per hunter per month, at the rate of 2.37
rounds of ammunition per animal.[61] In Hurungwe, the .303 hunters used
2.59 rounds per animal at an average of 4.38 animals per hunter per
month.[62] A total of 1,101 animals were killed in the Hurungwe West operation of 1959–1960 at 2.44 animals per hunter per month, with 3.48 rounds
per animal. In Hurungwe North and East, 1.385 animals were killed in
1959–1960 at 4.38 animals per hunter per month, or 2.59 rounds per
animal.[63]

These figures represent the intersection of the meat imperatives of
African hunters and the state's program for science and policy. As an illus-
tration of the duality of the transient workspace producing meat and
science for African huntsmen and the tsetse department, this intersection
requires a detailed examination.

Tsetse Control as Workplace, as Seen from the Village

Kubasa is the name vaShona give to a place (*nzvimbo*) where work (*basa*) is done. *Ku-* is a locative placeholder for *nzvimbo*. *Kutsetse* (*ku-* + tsetse) is the name they gave to the place where tsetse work was done. *Kutsetse* was the venue of *basa retsetse*, chiShona for "tsetse work." Like all matters of life, *kutsetse* and *basa retsetse* were guided mobilities. Moreover, tsetse work was hunting, which we examined in preceding chapters as mobilities in sacred forests that required spiritual guidance for forest animals to be revealed, pacified, and killable.

Just because it took place in a forest does not mean *basa retsetse* was unique compared to other workplaces. Since European occupation in 1890, there was now one big *sango* (forest) called *kuvangerengere*, the place of those who make strange, unintelligible *ngere-ngere-ngere* sounds when speaking (i.e. white men). This forest was not new; people had gone to *Natali* (Natal), *Kimbali* (Kimberley), and then *Joni* (Johannesburg) before to hunt and bring back *svexilungwini* (things of the white man's world).

Like all forests, this one required hunters to spiritually arm themselves prior to the start of the hunt. In the European conception, the forest was just a tsetse-infested space of trees and forest animals. African hunters did not embrace this mundane view; the forest remained sacred. Even in his Annual Report for 1952–1953, Chief Entomologist J. K. Chorley admitted that Africans regarded certain areas as "*haunted* by spirits [and] no native hunters can be persuaded to stay alone in these areas." As a result such areas were "inadequately covered."[64] There were taboos to be respected, including not defiling specific groves where the ancestors were interred, entry to which had consequences to the entrant, his entire family, and even an entire clan.

One will not find in the state records and annual reports any mention of the prayers and procedures each hunter conducted prior to reporting for duty. Interviews and experiential understandings of chiTshangana and chiShona will show the hunter appearing before the ancestors to ask that their path ahead be cleared of *zvimhingamupinyi* (obstacles) (Manyimbiri 1983, 134; Gwete 1983, 192). Just like before, the child of the ancestors informed them he was wandering into the forests (*kutetereka nemasango*), but this time looking for a job *kuvangerengere*. Beer was brewed and presented before the ancestors, that they might lubricate their thirsty throats, soften their hearts, and energize themselves to clear the path and forest trails ahead, guiding, and opening doors of opportunity (Chimera 1983, 174). Yet the pathways would only be safe and fruitful if the hunter and his wife kept taboos discussed earlier. Those taboos would be kept regardless of what *valungu* said about their business with the tsetse fly.[65]

The itinerations in *vangerengere*'s world were also hunts not only because they were menial, but also because one had to move from one place to another looking for work (*kutsvaga basa*), to advertise oneself in person. Workplaces were transient abodes lasting only until a friend or relative told of a better-paying *mulungu* or *murungu* (chiTshangana and chiShona respectively for white man), or one got fed up, or crop planting or harvest time beckoned, or the *baas* (the boss) was too cruel.[66] Hanyani Chauke (born 1923) is an example of this transience: he worked at Triangle sugar company for three years (1945–1948) before going to the copper mines of Musina. The mine owners then selected him as part of the team that surveyed and opened Mukondo Mine in Save under a white man named Fleming. Afterward Chauke left to go and work in the survey department of Mhangura Mine.[67]

Chauke remembers well the year when he left home via Mupfichani to Musina. It was 1948 when he walked this route for eight days; he remembers the year because *"nhiko ndinganyanya kuhlupeka la"* (that was when I was struggling so much).[68] At Musina he worked as a *muchini-bhoyi* (machine boy, the one who operates the machine for drilling) and as a *malayitsha* (the one who shovels ore into the cart).

He brought back *sithlatula* (shoes), *tihembe* (clothes), *mabande* (belts), and *xikanyakanya* (a bicycle). From Musina, the men of the village rode their *xikanyakanya* to Manyoni, to Hlengani, to Manyekanyeka, to Halata, to Makanyani, to Chikombedzi, to Gulungwini, to Machinzu, to Chingele, to Chibwedziva, to Bhaule, to Mfichani—nine days in total.

As a *magayisa* (a wayfarer who had been guided by the ancestors and now came back laden with goods) arriving from Joni, Gazeni Komundela and his peers were treated just as the *maphisa* of old had been treated: with ululations and clan praises, and much joy. Fathers and mothers would recite prayers to the ancestors, the whole line of ancestors, telling them that the children had returned from Joni, had been well guided in their itineraries, and now they were coming back home safe and sound, carrying *svexilungwini*. Komundela remembers well his father Chikambachi talking to his ancestors: *"Vabuyile vana lava. . . . Nikensile ngofu vana vangafika vayaJoni. Vabuyile vana vaminawo"* (Your children have returned. . . . I thank you very much that the children have returned safely from Johannesburg. My children have come oh).[69] To not say "thank you" was the worst form of witchcraft (Shumba 1983, 180). Without the guiding, steadying, and reassuring hand of *midzimu* in the lands of *vatorwa* (strangers), the itinerant

Figure 5.1
Gazeni Komundela.
Source: BBP

might have succumbed to indiscipline and returned with nothing (Vhuri-nosara 1983, 181–182). Hence the adage *natsa kwawabva, kwaunoenda husiku* (leave in good standing, for the path ahead is dark); one can always turn back to a loving mother, wife, and community.

Magayisa would bring things for everybody—an overcoat for grandma and grandpa; shoes and clothes for sons, daughters, nephews, nieces, and grandchildren; sugar, petroleum jelly, and soap for his wife and children; money to buy cattle for himself and for his parents, and so on.[70]

Whatever the *basa* one was involved in, one was never to claim personal credit from what was after all guided mobilities. These hunters were to remember that ancestors were the sources of vision and hearing, givers of receptive yet adjudicative minds, fountains of all blessings. The loss of or contempt for such guidance rendered *vangerengere's* knowledge a kind of madness, the itinerant being lost and wandering in the dark forests of *vangerengere's* world *sebenzi* (like a madman). For only the lunatic could say, after finding things by way of guidance: *Ndazviwanira ungwaru hwangu nemano angu* (I have discovered wisdom through my own abilities and hard work). To be ungrateful is, after all, more than madness.

Unlike the *magayisa* of the nineteenth century who walked from the last train station in South Africa, the generation of the 1940s and 1950s rode on their "28 Humber type" manly bicycles identifiable by their crossbars and structural strength. They purchased them for the work of patriarchy, principally to do things that would sustain the family and enable mobilities in a forested place full of obstacles and opportunities—like hunting in Gonarezhou. The 28 Humber was called *bhasikoro remutanda* (bicycle with a crossbar). Women could not ride it because the pedaling up and down lifted their skirts.

The bicycles had been purchased in *Joni* (Johannesburg, where local men worked intermittently). They were put on a train to Beitbridge or Louis Trichardt (Polokwane), from where the men of the village offloaded them, then secured their parcels onto bike carriers, crossbars, and handle bars and cycled home via Bendemutale or Makuleke, crossing the Limpopo into Rhodesia. Each bicycle cost five bob (sixty cents), two months' wage in the 1950s and 1960s. The men from Joni usually worked a full twelve-month contract before getting on the road to *kanya-kanya-kanya-kanya* (pedal) home.

To own a bicycle meant one was recognized as a real man. The only other men who could afford *xikanya-kanya* were farmworkers on white farms, hunters, and veterans, who had received bicycles as reward for fighting to save the British Empire in World War II. (By contrast, their white counterparts got mineral concessions and large commercial farms from rich land seized from Africans, who were forcibly resettled in areas cleared of tsetse, but prone to reinvasion). When bicycles broke down, they were repaired locally with spares from South Africa brought in through the same routes as the bicycles.[71]

As noted, the ancestrally guided man was one who returned home "carrying something in his hands." To carry something (*kugukuchira*) was also the sign of a successful hunt, an occasion for women to ululate, dance about, and recite prayers (praise poetry). While away in Johannesburg, Musina, Mhangura, or Triangle, in *vangerengere* or *valungu's* world, these men would consume certain foods and see how certain machines worked, how *valungu* built their houses, how fruit trees grew in their orchards, the "gum trees" (eucalyptus trees) and vegetable plants *valungu* grew, how they plowed the land, and how they entertained and kept themselves informed using *xiwailesi* (wireless, or radio).

When they got their wages, Africans purchased some of these *svexilung-wini*, and brought them home with them, to plant them in their own homesteads, to build their own versions of such houses, to replace hoes with plows, and so on. Africans based in *valungu*'s world amassed large herds of cattle, built tin-roofed four-cornered houses with fired (red) bricks plastered and painted over and bought and installed hammermills at grocery stores (*magirosa*) to grind grain into mealie-meal as well as butcheries to sell beef, now that animals were depleted in most places.[72] The village, therefore, is the perfect place from which *kutsetse* and *basa retsetse* will now be analyzed, not according to the state's reasons or mine, but these hunters' own.

Why Did Africans Enlist as *Magocha*?

They became known as *magocha* (singular, *mugocha*): those who are always roasting meat. They were men whose access to meat came to define their identity category in society. Men for whom *basa retsetse* could only mean something—anything!—if it delivered one vital community for family, kin, and as an exchange commodity with which to access means to exercise a modernity defined by themselves under *vangerengere*'s rule.

In a time when Africans had been deprived of meat through the criminalization of the hunt and severe limits to cattle herd size that turned any slaughters into a luxury, eating meat became a measure of one's modernity. To eat meat was to be like *valungu*; for meat could only be bought in butcheries (*siraha*) at *magirosa* or in town, which made it *svexilungwini*. Only *valungu* (black people who had means *to be white*, to eat, dress, and live like whites) could afford to buy meat. To be married by a man who was a *mulungu* was to also become one, hence the willingness for women (and parents) to marry (their daughters) into polygamous unions. "People bought cattle from the Ndebele with money from that *gotsha* work," remembered one chief. "It seemed so much money, you could buy quantities of goods. . . . Some of the *amagotsha* accumulated many wives" (Chief Kavula, cited in Alexander, McGregor, and Ranger 2000, 41).

Thus even when the Department of Tsetse and Trypanosomiasis Control introduced a bonus for hunters to kill animals that were inedible and whose skins had no saleable value, African hunters ignored the instruction. They calculated instead to target those animals that made far more capital—social, religious, financial, or otherwise—than the few pence and shillings the department offered. The officials identified the problem correctly: the hunters utilized all the edible meat of the animals they shot, and were naturally inclined to satisfy their needs before expending their limited ammunition on *worthless* animals like baboons. Hence, any reward for killing them would increase the death toll "only if the [hunters] otherwise destroyed more edible game than they could utilise."[73] Because the principal attraction of a hunter's job *was* "in the meat obtained," the chief entomologist made it clear that "if his rights to it were lessened we would simply have no hunters."[74] The hunters had one advantage: these were borderlands and they were local residents on the margins, over whom government control ranged from tenuous to nonexistent.

John Piet Barnard was a *mugocha* at the height of the anti-tsetse operations in the Save and Lundi valleys from about 1952 onward. He was the son of Bvekenya, schooled in the finest traditions of the hunt by his father's maTshangana hunting buddies, now that Bvekenya himself had retired to a life of farming in South Africa. Department of Tsetse and Trypanosomiasis Control officials got wind of John Piet's exploits at Lone Star Ranch, where owner Ray Sparrow employed him to destroy vermin (lions, hyenas, wild dogs, and leopards) terrorizing his livestock along the Lundi River. During weekends and holidays, Sparrow allowed John Piet to carry the .303 rifle he used for his job to hunt for his family.[75] The Department of Tsetse and Trypanosomiasis Control had just switched from the Martini-Henry to the .303 rifle when officials recruited John Piet, and let him keep his rifle at home so long as he submitted the tails of animals he killed each month to the tsetse ranger at Chivonja Hills, near Chipinda.[76]

To appreciate how men like John Piet extended to *basa retsetse* their knowledge of the forest acquired from the local professoriate of the hunt, we must probe a bit further into his personal life. He straddled two spiritual worlds as a *n'anga* (healer) and a Christian who went to church every weekend. He allowed his own children to go to church, but the church did not replace or address ancestral spirituality. In those situations, John Piet put his snuff down, brewed beer, and summoned his *shavé*, Dodorolo Kufakuzvida.

Dodorolo was not only *shavé rekurapa* (divinatory spirit) but also *shavé rekuvhima* (hunting spirit). As *shavé rekurapa*, Dodorolo drew to John Piet people from near and far, from Rhodesia and Mozambique alike, coming to get some water to bathe and chase away their *minyama* [bad luck]. His shrine was a pit two paces wide, to which his wife Nwa Jeke brought a pot of boiling water containing mixtures of medicines. John Piet then put the steaming pot at the base of the pit, then asked the patient to go on all fours, facing down into the steam with a blanket covering him. Meanwhile, outside the blanket, John Piet made cross signs while barking at the bad spirits to come out of the stricken patient in the name of Dodorolo.[77]

The space of medical practice was located outside the homestead under the *muhacha*, the tree of the ancestral spirits. The site was usually in the nearby forest, so that when the pests came out of the body they went straight into the forest, which as we saw was the primary dumping ground for wandering spirits and pestilence. Moreover the patient came out of the blanket in a stupidified, even disoriented state, sweating, sometimes even having passed out, and needing *kufurwa nemhepo* (fresh air). John Piet had learned from his mother, Nwa Muchisi, that the smell and steam from boiling water filled with the crushed bark, roots, and leaves of the *chinwamaruka* tree was so powerful that any unwanted spirits would leave the body alone and return to wherever they had come from.[78]

Shavé rekurapa brought John Piet substantial wealth. He did not charge a fee and preferred that patients only express their gratitude after they had fully recovered. They would then come to tell John Piet to send his sons to *avuya tateka homu netimbuki* (to come and collect a cow or a goat) because the patient had been cured.[79]

John Piet also had a *shavé rekuvhima*. Together with his brother-in-law, Gazeni Komundela, son of Bvekenya's hunting mate Chikambachi Komundela, John Piet would go before their ancestors to ask for guidance in the hunt. Both of them respected the abstinence from sex and other rituals. When he killed, John Piet would kneel down and say: "*Matahe Matahan'ombe hakensa, mukahe nika nyama, Dodorolokufakuzvida*" (Matahe Matahan'ombe thank you, you have given us meat, Dodorolokufakuzvida). As he uttered those words, he would take the liver, throw it in pieces to the four sides of the forest, then take the carcass and go home.[80]

Valungu did not consider the ancestral spirits as agents in the skills of *magocha* like John Piet. As two senior officials observed, African hunters were "generally good *veldmen* [bush experts] and trackers but poor shots and their shooting was not improved by the worn state of the ex-military .303 rifles with which they were issued. A 90 cm grouping [on the target board] at 25 metres was considered reasonable and few claimed to be able to hit a 30 cm diameter tree at 100 paces with confidence, whilst most did not attempt to shoot, even large antelope, at ranges of over 150 metres."[81] These comments reiterate the misconceptions of Selous and others (which Parker Gillmore and John Millais corrected in chapter 2): to measure skill according to how accurately the hunters shot as opposed to the skill of maneuver. These value judgments could also be politically charged (Storey

2008) and intended to advance a specific program—in this instance to push for getting rid of independent-minded, meat-seeking African hunters and replace them with department-trained hunters subjected to drill and military discipline.

This independent-mindedness was on display in the southeastern fly area two years after John Piet began *basa retsetse*. His son Piet Barnard was fifteen and learned the art of hunting by accompanying his father on these shooting operations. Piet Barnard makes it clear that his father treated *basa retsetse* no differently from hunting for the family or for sale: the overriding imperative was meat, skins, and ivory. He generally paid no attention to what the fly ranger, who was at [Save] Camp, said and had no way of enforcing.[82] In 1954, for example, the department lamented that local hunters did not "hunt where directed," especially if the area was too far away from home.[83]

How Did the Meat Imperative Shape *Basa reTsetse*?

Entomologist Rawdon Goodier detailed the importance of the African hunter's meat imperatives in the actual conduct of tsetse operations in 1956:

There is little doubt that the hunter's first desire is to provide meat for himself and his dependents. In the Chirundu area, where there is not a large potential market for the meat, the hunter's requirements over the years would remain fairly constant. In the early years of the game destruction operations, when game was abundant in the area, his meat requirements could be satisfied with very little effort, so that the . time spent hunting was probably quite small. As the game destruction operations progressed, the game would become scarcer and harder to shoot, but his requirements being the same, the hunter would have to spend more time hunting to obtain the same kill. In this way the monthly game totals would remain constant for a considerable period in spite of changes in the density of the game. The hunter also has preferences for the type of game he shoots—a small buck such as an impala, or duiker, requires practically as much time and effort to shoot as a large one, like a kudu; but the meat return is much smaller, therefore, when game of all kinds is plentiful, large buck are likely to be shot in preference to small, the small ones only being shot when the large become scarce.[84]

The kill rate depended not only on the numbers of hunters but also "their meat requirements." It was common for hunters to "shoot two or three small animals early in the month, so as to justify their employment and meet their immediate food requirements in the field, and one large animal towards the end of the period so as to have plenty of meat to sell," especially just prior to Christmas. In the Chiware area of Sebungwe, it was therefore resolved in December 1959 to dismiss the hunters home "as soon as each camp fulfilled its assessed quota based on past performance," that is, after eight to nine days instead of the usual twenty-five days.[85]

The more animals they killed, the more difficult it became for African hunters to find meat. So they identified areas outside but near their designated operational area, then, made false reports that they had caught tsetse fly there. The European tsetse rangers trusted the hunters as "experts at catching tsetses," but by 1949 had become aware the hunters were making "false reports" to seduce their white superiors into extending the shooting line. The hunters also deployed this trick whenever it was certain that game elimination had succeeded in an area and the operations were winding down, whereupon they would have to surrender their rifles and go home.[86]

It took until 1955 for the Department of Tsetse and Trypanosomiasis Control to find out that the hunters on *basa retsetse* were supplying ammunition to their colleagues in the villages engaged in "poaching." In return, the latter supplied the hunters on tsetse duty with tails to create what one entomologist called a *"tail bank"*—a surplus of tails the hunter on tsetse duty could declare as evidence of kills at the tsetse ranger's camp every month end. Africans living in designated Native Reserves, Special Native Areas, and Native Purchase Areas were allowed to destroy animals in their localities, but they often hunted further than the state permitted. A family member on tsetse duty usually asked others in his family to spare the tails of all animals they killed so that he could declare them as his own kills, adding them to his tail bank, and thus retain the ammunition he didn't use for those kills for his own personal use.[87] At the end of each month, every *mugocha* submitted tails of all animals killed, and had his permit stamped to confirm his returns. He simply tapped into this tail bank when it was time to submit his monthly returns to the tsetse ranger.[88] The *magocha* picked ammunition rations and returned to the "battlefield," or straight home to do their own personal work in the village, returning only for the imperative of meat, or to pick up tails from their "bank" for submission to the ranger. This is how African hunters were able to build a strategic reserve of ammunition, first for their Martini-Henry rifles (up to 1952), and then .303s thereafter, and continue hunting until gun control and antipoaching regulations and patrols tightened in the 1960s.

The government became completely powerless to stop *magocha* from killing for meat under the alibi of *basa retsetse*. Minutes of one local (European) conservation organization express a sense of utter resignation to the African hunters' subversion of the entire purpose of game elimination:

"It is impossible for control to be maintained over Native Hunters. . . . The system whereby the hunter is paid only £1 per month and allowed free sale of meat etc. derived from game shot as an incentive to holding the job is unsatisfactory. . . . There remains literally no game within the shooting belt, with the result that hunters obtain their incentive over a wide area. . . . In [as] remote a region as the one in question control is difficult."[89] The assessment of the director of tsetse control operations in 1956 was also quite grim: "There is little or no indication that the game destruction process was having any effect whatsoever on the game population in the area, the hunters being compared to a very small and constant number of predators in a large population of potential prey. They may be likened in this respect to most carnivorous predators which do not, under normal circumstances, cause a reduction in the prey population, but are merely one of the many factors which prevent an infinite increase in the population of their prey."[90]

The blame was put squarely on the tsetse ranger. As was standard colonial practice, every government activity required the supervision of a white man. The tsetse rangers who were supposed to supervise African hunters were located at camps by the roadside, far from the margins of the colony. They had no vehicles—let alone lorries or 4 × 4s; even if they did, thick thorn trees dominated the roadless terrain where the actual shooting was happening.[91] The rangers would visit the nearest store "once a month to obtain monthly supplies of food, etc." but the life of a tsetse ranger in the forest was lonely; once he went to town he seldom came back. Fewer and fewer Europeans enlisted.[92] Of those that did, "half . . . proved to be unreliable, most . . . stayed for only two or three months and either resigned or [were] dismissed"; otherwise, for most of the time there was no *mulungu* to supervise *magocha*.[93] In other cases the *valungu* were simply too lazy to even leave their camps.[94]

Innovation Forced from Below

The professionalization of game elimination was an innovation forced from below. In 1971, Chief Glossinologist Desmond F. Lovemore concluded that "hunting operations were fast settling into a rut, in fact in some areas this had already occurred."[95] Game elimination was fast becoming "literally nothing more than the reporting each month, in our various monthly reports, that so many animals were destroyed in this or that hunting section or operations area, with no further attention being paid to the work."[96] The work of starving the fly, Lovemore stressed, could only succeed if every person in the Tsetse Branch, top to bottom, was synchronized into one mobile machine raging against the insect, "constantly stimulated to approach their task intelligently and with enthusiasm."[97] Instead, the whole thing had turned into a meat-producing machine for villagers.

How then to liberate the hunting operation out of the domain of meat production for Africans to that of starving and immobilizing the fly? Entomologist Ted Davison, whom Lovemore assigned to go and assess the performance of hunters in the field, identified two key problems. The first was that of the "long serving reliable officer becoming infirm" and *just sitting there*, waiting for tails from African hunters and then compiling reports. The problem was that the earlier generation of hunters already schooled in the indigenous professoriate of the hunt, on whom the tsetse ranger's job had relied, was thinning out of the system in the 1960s.[98]

Davison did not say that the state was responsible for this: as game reserves were established and the professoriate of the hunt was criminalized as poaching, in most areas the venue to teach and learn in, indeed the material to practice with, was thrust beyond the reach of the old-timer hunter and his trainees. British-style schools and other distractions did not help in grooming new hunters either. With that came also a deficit in skills. Davison lamented that the department would have to embark on training hunters to shoot, track, and hunt, something that fathers had taught to their children. In essence, the tsetse ranger was now called upon to be the new professor of the hunt, the recruits coming into tsetse service raw. Training was going to require books, pens, blackboards, and chalk, with a curriculum of writing, practice sessions shooting on the range, and bush craft to apply in the field what was taught in class; certificates were to be issued after successful completion.[99]

Training aside, the hunting operation was also in urgent need of water-tight supervision and central coordination if it was to redirect the hunters' energies from supplying meat to the village to advancing science and policy. Accordingly, all hunters were from 1958 onward required to be based at fixed camps that the tsetse rangers could access by road. Deployed in rotation in groups of two or three, the hunters were assigned to "fixed camps located so as to achieve optimum coverage of the area, taking into account places that were particularly attractive to animals."[100] At night the hunters slept in these semi-permanent camps, hunting from dawn to noon, when they broke for lunch before hunting again from 2 pm to dusk.[101]

After the moratorium on hunting between 1961 and 1964, the original system whereby hunters were dispersed in groups of three was again changed. This time twenty hunters operated from one camp under a (white) tsetse field officer (TFO, formally tsetse ranger), returning there every night. A senior tsetse field officer (STFO) administered a cluster of these twenty-man teams. Every hunter was supposed to "account for every shot fired" to his TFO; in turn, the TFO submitted a monthly return to the STFO detailing the number of animals killed and shots expended on each. This enabled the field officers to determine the average number of shots fired per animal killed, "as a measure of the efficiency of the hunting teams."[102]

Besides training and centralized control, a third move toward professionalization was the introduction of a salary and other incentives. This move had already begun in 1955 with Chorley's request to the Ministry of Agriculture for "the substitution of paid native hunters for unpaid native hunters" in the Chipinge and the Save-Lundi areas. The reason, he said, was that this "would enable the Ranger i/c [in charge] to exercise greater control over the hunters and ensure a better distribution of the rifles."[103] In a follow up, in 1970 a bonus scheme was introduced with two aims. One was to "increase the welfare of the long serving hunters" to enable them to stay and thus avoid having an inexperienced workforce. The other was to incentivize "all hunters to put more effort into the[ir] task."[104]

Now that the meat incentive had diminished because of the stricter control and ammunition count, the success or failure of game elimination rested on a good salary and an incentive scheme to make up for the losses.[105] However, as Davison discovered the bonus scheme had "not provided the desirable increase in the hunters welfare," while the hunters found the earnings so worthless and accessible to so few that the job was not worth taking up. After the initial propaganda regarding its virtues, the effect quickly wore off and "even some dissention has arisen due to the unequal chances of earning bonus amongst the teams."[106] On paper, each hunter was supposed to earn R$5,9 per year as an incentive, which was already laughable, but they ended up getting no more than R$2—if they got even that. Davison was blunt: "The scheme does not provide any incentive to the majority of our hunters."[107] The bonus scheme was discontinued in 1972–1973.

Conclusion

Writing in 1955, Aimé Césaire described the tendency of colonialism "to drain [Africans] of their essence, trample their culture underfoot, undermine their institutions, confiscate their lands, smash their religions, destroy their magnificent artistic creations, and wipe out their extraordinary *possibilities"* (Césaire 1955, 42–43). Indeed, Césaire summarized the victimizing tendencies of colonialism rather well:

They dazzle me with the tonnage of cotton or cocoa that has been exported, the acreage that has been planted with olive trees or grapevines.

I am talking about natural economies that have been disrupted—harmonious and viable economies adapted to the indigenous population—about food crops destroyed, malnutrition permanently introduced, agricultural development oriented solely toward the benefit of the metropolitan countries; about the looting of products, the looting of raw materials. (Césaire 1955, 42–43)

Yet underneath these undeniable victimizing tendencies, the mobilities of ordinary people and the work they do suggests a different narrative in which *the colonized built the colonial system.* Colonial tsetse policy was shaped from below in a very fundamental way. Because *magocha* operated virtually on their own (singly or in groups), they acquired such latitude as to determine every outcome of game elimination—the statistics the Department of Tsetse and Trypanosomiasis Control produced, the conclusions officials made of them, and the policies and "scientific" decisions arising from such conclusions. European tsetse fly rangers who were supposed to supervise the African hunters were located at camps far from the hunters and most never stayed on the job long enough. That meant the only chain of consistency tying the tsetse operations together was *mugocha.*

The hunters were not here for the subtleties of science and policy, far from it. They were here for the meat. Yet, in killing for the meat they also killed for science and policy. The government cared less for the nutritional needs—let alone welfare—of the African. Yet, in deferring to the *vahloti* and availing them the space, animals, and guns, it provided them the instruments to become *magocha*, a rich vein of meat supply to the village. Every animal shot for meat was, in any case, one less potential carrier of tsetse fly. That too was good policy.

Meat requirements determined the *mugocha's* "kill rate," enthusiasm, and whether or not he stayed on. This kill rate was also a function of the material properties of the guns *magocha* used, a factor clearly illustrated when the Tsetse Branch made the changeover from Martini-Henry to .303 rifles in 1952, citing dwindling stocks of ammunition for and the extreme state of wear and tear of the former rifle. The kill rate worsened, illustrating the difficulties of adjusting to new technologies. Materialities notwithstanding, Africans could falsify how many rounds they had actually expended on animals, and then use the ammo for illicit ivory hunting.

Rather belatedly, having been defined by its absence—hence the reliance on African hunters as frontline soldiers against invasive species—the state now sought to define itself by its presence. In an effort to seize the initiative and end the bottom-up agency of Africans over tsetse operations, from 1958 a more centralized system was introduced in which all *magocha* were co-located with the tsetse ranger, who was now rigorously trained to enhance his control of hunting teams. The professionalization of the Tsetse Branch's field operations was forced from below.

Therefore, revisiting even those spaces that are often marked as Europe's civilizing missions—that are attributed to top-down, global North–global South technology transfers—one finds that few if any Europeans actually did the dirty work of chasing after elephant. The *nzira* (ways and means) required to transform the colonizer's "grand plan" to create farms, establish disease-free spaces, and build roads and cities was not necessarily drawn from the European's order of knowledge. True—the plan of the building might be from London or Manchester, but the means and ways of turning the white man's designs into a physical reality could actually be based on resources endogenous to the societies of the colonial subjects. The initiative to apply those skills might, as in this chapter, have nothing at all to do with the altruistic desire to see science succeed, but with the more pragmatic and mundane necessity to fashion *nzira* (means) of everyday life. This matters even more so in a situation where conventional means of survival were criminalized as poaching.

6 Poaching as Criminalized Innovation

It is important to appreciate that, within the African community, there is absolutely no resistance to the snarer and nobody, from the Chief through the school teachers right down to the simple peasant, ostracizes a man who cruelly ensnares wild animals regularly. No one ever thinks of reproaching him. The snarer is recognized as a man of standing in the community—a supplier of protein just as the old English rabbit poacher was regarded by the inhabitants of villages in that country in the days of large estates and wealthy land-owning squires.

The African snarer's activities, far from being discouraged, are actively applauded by every man, woman and child in the neighbourhood. Is there nobody who realizes that indiscriminate snaring and hunting must eventually mean that there will be no game left to trap and is not a voice heard anywhere protesting about the cruelties of the trade? The answer to both questions is NO. The African lives only for this day and cares not what the morrow will bring—he is by nature an absolute fatalist who makes no attempt to alter courses which the spirits have probably already planned for him. As for the cruelty part, the average black African has no feeling whatever for animals—to him there is as little wrong with chopping down a tree as breaking an ox's leg or skinning a goat alive to obtain a pair of leather bellows—trees and animals are all "things" which have no feelings at all and merit no consideration whatever. All of us who have the welfare of dumb creatures at heart, must wonder when the African will develop some compassion towards animals. I personally believe that this may never come about—after all many white races that have long centuries of "civilisation" behind them, have still acquired no feeling at all for members of the animal kingdom. . . .

One of their senior headmen, on being asked by me about the snaring activities of his people, replied: "Can I stop my followers from drinking water?"

These are the words of former district commissioner for Nuanetsi (now Chiredzi) Allan Wright, writing in his memoir, *Valley of the Ironwoods* (1972, 142–143). He was summing up something that appeared puzzling: that "barbarians" destroying nature instead of conserving it could be popular in society. The irony of the colonial project that Wright oversaw

Figure 6.1
Chibwechitedza, District Commissioner Allan Wright.
Source: NAZ

on the ground is how convenient it was to criminalize the professoriate of the hunt and yet rely so thoroughly upon it as a technology of tsetse control.

While in Wright's eyes and indeed in conservation, tourism, and state circles the poacher (*mupocha*) was seen as a criminal, in the villages *mupocha* was—and still is—a hero. In fact, only the *majoni-joni* (as local maTshangana call those who work in *Joni* or Johannesburg) is as popular. Locally the colonial language of criminalization is derogatory and in breach of history; it is also interpreted as the *valungu*'s (white man's) acknowledgment of maTshangana and vaShona's *unyanzvi/vutivi* of hunting; poacher is another way of saying *hombarume* or *maphisa*, scornful and unjustly put, but an acknowledgment of indigenous expertise nonetheless.[1]

Very well then, *valungu* could install fences to keep the animals away from them using the excuse of keeping tsetse fly and tsetse-carrying animals out. They could deploy guards and patrols. *Valungu* could take away their guns. They could come, arrest them, and send them to jail for hunting.

But there were ways around that. The fences were not impregnable; in fact, they made good wire for snaring animals inside the park itself. The guards were close kin—uncles, brothers, nephews. They would guard, watch, and follow the movements of the animals and relay the message to their kin; when patrolling with untrusted outsiders (those recruited from elsewhere) and *mulungu*, they would warn their kin to lie low or go in the opposite direction. And the record of the colonial state in disarming them was not that impressive anyway.

Nor was the prosecution, which, assuming it was successful resulted in a few days in the *trongo* (tronk, or prison). Otherwise the hunters could always pay a fine, which money they kept ready just in case the prosecution's case was successful. Or they could just send juveniles to hunt or collect carcasses, knowing they would at most be canned, and then be released because they were underage under *mulungu*'s laws. In fact the consequences inscribed a layer of heroism to poaching.

Bvekenya Dhulu laMahlengwe

The heroism of the poacher transcended race. Already in chapter 3 we saw the solidarity local hunters built with Europeans coming to hunt in Gonarezhou and the adjacent parts of Mozambique. The beginnings of the criminalization of the hunt are located in this early period, which shows what happens when government alienates locals, by taking away their land and giving it to animals and white tourists, leaving them with no choice but to form alliances with incoming actors that contest this state authority and monopoly over this "wildlife." This is a familiar story in Africa, namely, of the alliances that form between local hunting and forest practices and the activities of international ivory poachers and dealers, or, at the very least, why locals end up not being of assistance to the state and park in shutting out incoming commercial actors.

Creating Alliances through "Marriages"
The circumstances within which a *bhunu* (singular white man in chiShona) named Cecil Barnard became *Bvekenya* and a husband to three maTshangana women illustrate how crucial it was for *valungu* coming to hunt outside the law to establish alliances with local villagers quickly. That locals entered such partnerships to start with is an indication of their self-interest. Bvekenya got his name after a thorough beating from local maTshangana highwaymen in the Rio Save area of Mozambique in 1910 that left him with no clothes on his body. As he trudged back to Makuleke, the blisters

Figure 6.2
Cecil *Bvekenya* Barnard.
Source: Solomon Bvekenya

on his feet were so painful that he was limping. Along the pathway he encountered a group of maTshangana men returning from the Rand gold mines, who upon observing his unusual walking behavior called him *Bvekenya* (the-one-who-walks-with-a-stagger) (Bulpin 1954, 205–206).

Valungu's deference to local culture as a means of making allies necessary to gain access to the resources of forests Africans believed to be sacred opens up an interesting dialogue on culture as technology. Hence to cement their relationships with local villagers, *valungu* married local women under customary law. In Rhodesia, Bvekenya went on to father children with at least three women: here two of them are central to this chapter: Nwa Muchisi in headman Masivamele's, and Nwa Makavhela of Chief Mahenye's.[2]

When coming to Masivamele, Bvekenya used to arrive at the home of his friend *Juvendava Muguri weMhosva* (Juvendava settler of disputes), who was otherwise also called Tshihosi. In the course of coming to visit Juvendava, Bvekenya soon laid his eyes on a woman named Nwa Muchisi, daughter of Hlopekazi, who stayed close by. The friendship with Juvendava was *chihwande* (ploy) to gain entry into the village so that local hunters could then show him areas where elephant was to be found. Juvendava

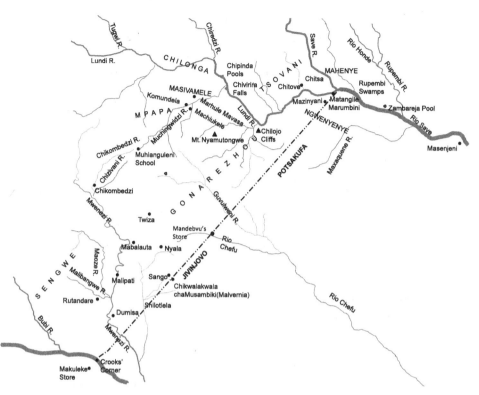

Figure 6.3
Map of Gonarezhou and its surroundings during Bvekenya's time.
Source: Author

was the one who knew *"nzvimbo dzaishandiwa kuno uku"* (workplaces around here) that is, hunting grounds.[3]

Thus began Bvekenya's secret liaison with Nwa Muchisi. In the strict chiTshangana sense, he did not marry Nwa Muchisi; in practice he did because he left her father lots of cattle. It was the white man's way of paying *lobola* (dowry) and how *valungu* showed their appreciation. MaTshangana defined a "proper" marriage as sitting down and doing every step of the rituals, with the whole village and extended families invited, each item demanded by the father- and mother-in-law being counted and given by a *munyai* (go-between), with the women cooking *mafetikuku* (fat cooks, or pancakes). However, Bvekenya went directly and quietly, paid the bride wealth to his paramour's parents and the whole transaction was finished.[4]

Men like Juvendava were important nodes of information and security; nothing that was important escaped his ears. He also organized the logistic resources for white men coming into the area to hunt. One of them was tracking—Juvendava knew exactly who Bvekenya might need within Bhaule area, further south in Malipati area, or east in Mahenye. The other service was cooking food and washing clothes for *valungu* like Bvekenya. In chiTshangana custom, fathers could and still can avail space for young men who seek to speak to their grown up daughters, so long as they show respect and observe protocol. So there was nothing unusual in parents sending their daughters of marriageable age to "cook for" Bvekenya. Not only did they bring back meat; Nwa Muchisi in particular became pregnant, and Bvekenya could not deny responsibility without poisoning relations critical to the hunt, so he made peace. The elders knew how to snare strangers who possessed resources with locally usable value. At that time, what counted as support for the children was buying them clothes to wear and bringing them meat from the hunt; on that measure, Bvekenya exercised the role of husband and father to his African family.[5]

Bvekenya also created another circle or node of "allies" or "trackers" in Chief Mahenye's area of Chipinge. As a young boy Gazeni Komundela (born 1930) saw Bvekenya on his many sojourns in Mahenye; his father Chikambachi Komundela sat on many occasions at *huvo yavanuna* (men's fireplace) with two of Bvekenya's hunter allies in Chitove: Matangile and Mazinyani. He recalls the poacher's two hunting camps at Chitove in Mahenye and Lisungwe in the Chinzine area.

Inevitably, just as in Chibwedziva, Bvekenya needed "help with cooking," "washing clothes," and general cleaning, and again, parents assigned their daughters of marriageable age to go and offer such services. This is how he "married" Nwa Makavhela Chitsine, his local wife; it was Matangile who engineered the affair, his reasoning being to cement the relationship with this *mulungu*, a reliable source of meat and ammunition for .303 rifles, as well as Martini Henry rifles and powder for *migigwa* [muskets].[6]

The layout of Bvekenya's hunting camps demonstrates the symbiosis between the professoriate of the hunt and these incoming white poachers. At Lisungwe he specialized in hunting *tihlengani* (genets), *tiingwe* (leopards), and other colorful cats widely available in the area. At Chitove he hunted *tindlofu* (elephants) for ivory and *mampungulu* (hippos) to make *mbhoma* (whips). His route south to Makuleke, his market, crossed Marumbini through the dry desert-like stretch called Potsakufa (where it is easy to die) to Mandebvu's store on the Chefu. From the Chefu, he stuck the path called Jivingovo to *Chikwalakwala chaMusambiki* (Chikwarakwara inside Mozambique) and thence to Papfure (Pafuri).[7]

Figure 6.4
Mavuyani Koteni Sumbani.
Source: BBP

In this particular case, the flexibility of belonging also meant that when migrating or visiting their kin, locals would also be extending those networks for Bvekenya's use. Mavuyani Koteni Sumbani remembers that his father Dick Sumbani and Bvekenya already knew each other by the time that he migrated from Mozambique to resettle in Chibwedziva. The two hunters met by chance when they were hunting elephant in the *simbiri* area of Gonarezhou.

Dick Sumbani took the *mulungu* into the village, where he introduced him: "*Lo uBvekenya, dhulu lamaHlengwe*/This is Bvekenya, the granary of the Hlengwe."[8] The two men had hunted together in Mozambique, killing elephant and harvesting ivory; after the kills they would send a message to villagers to come. The hunters would then move on, having harvested their ivory, while the whole village *vasala vacheka vaja nyama* (would remain behind skinning and eating meat). That is how Bvekenya got the honorific name of *dhulu lamaHlengwe*; he was the local people's granary. Sumbani not only introduced Bvekenya to his new neighborhood in Chibwedziva, he also tracked for him so that *wadhuvula ndlofu* (he shot elephant), took the ivory, and left *nyama* for his people.[9] Soon people in Chibwedziva began coming forward to report elephant sightings to him. When the big tuskers were finished in a given area, Bvekenya left for Chitove or further south in Malipati area. The locals would survive on the *chimukuyu* (dried meat) for many months to come.[10]

Bvekenya was not the only one to engage in intimate relations with local women; practically every white man on the frontier did. His hunting and recruiting partner Charles Diegel, a German, lived with a local woman in the Mupfichani area until his cook ran away with her.[11] Even Native Commissioner Peter Forrestall—the face and voice of the law—had several African wives (Wright 1972, 25). While the state actively discouraged these relationships, on the frontier those racial boundaries were fragile. Such liaisons anchored the white man in the village. Nor were the benefits of such liaisons one-sided. There was a distinct advantage in providing sexual and other services to men like Bvekenya: forest meat was assured at a time when Africans were now banned from hunting legally.

Bvekenya and Mubhulachi Mavasa

In the land of Bhaule is a place called Chitalahimbera (a pool that fills up after the lightest of rainfalls). An old man named Marhule Mavasa was the sitting *Sabhuku* (Headman) Masivamele when Cecil *Bvekenya* Barnard trudged into the homestead around 1926. Locals across the border in Portuguese territory had directed him to this place because elephants favored it. Big tuskers.

It was Old Mavasa's grandson, Julius Mugocha Mavasa, who related the story of Marhule's adventures with this strange white man, who lived with Africans at a time when such behavior was out of character with whiteness. Bvekenya brought with him his gun and explained to Old Mavasa that he could hunt if there were animals around. He asked the old man if he knew someone around young and agile but with knowledge of tracking. It was a drought year, and the old man was barely managing to survive through the industry of his son, Mubhulachi Mavasa, to whom he had bequeathed the art of hunting, now that his own flesh was tiring with age. Mubhulachi Mavasa's own story would be told via his son, Julius, one of the most famous *magocha* in Chibwedziva.

Old Mavasa referred Bvekenya to two people whom he felt could help him. One was an old man (younger than him) called "Fifteen," whose rather strange name is an example of how Native Commissioners erased the meaningful names African parents gave their children because they were too cumbersome to write on identity documents and simply gave

Figure 6.5
Mugocha Julius Mavasa recounts the story of his grandfather, Marhule, and his father Mubhulachi in 2011.
Source: BBP

them any that came quickest to their minds. Fifteen was also known locally by another name, Zhuwawo, a corruption of the Portuguese name João, which means John in English, traces of his stint in Mozambique, part of the flexible citizenship on the borderlands, whereby one's abode shifts according to opportunities or constraints. So Bvekenya employed Fifteen Zhuwawo and Mubhulachi Mavasa to lead him through the rivers and other thickets in search of animals, waterholes the animals favored, and the times they were most likely to come.

Thus began a union of a white man and Africans in hunting. It was a partnership based on two dialectics, Bvekenya's usefulness to Old Mavasa as a *mulungu* with access to unlimited ammunition and a more lethal gun, on one hand, and Old Mavasa's position as *sabhuku* respected by his subjects, opening the door to a whole reservoir of the professoriate of the hunt, embodied in his own son, on the other. Incoming and endogenous became one, traveling, tracking, camping, teaching each other, co-innovating on the move.[12]

Fifteen was a very good marksman, who had been trained to shoot by his father, who in turn had been trained by his father. But now he was too old for the amount of energy required to track for days, weeks, and months on end. Mubhulachi had not fired a gun when he met Bvekenya, but he was young, full of energy, and an exceptional tracker who had apprenticed with his father. Bvekenya chose youth and energy over experience. So the two went away, Mubhulachi tracking, Bvekenya shooting.

Initially Bvekenya was interested in finding animals—*any animal*—so that he could kill and bring meat to the village of Mavasa, to thank this man who had opened up to him. In turn, Mubhulachi's father Marhule appreciated the human spirit in this *mulungu*, who had found it worthy to live in his own household, as if his own child, and hunt for his new-found family. At that time there was a serious problem not only of *nyama* (meat) but *usavi* (relish) in general. Guns might have been around but ammunition was a problem. If his son Mubhulachi did not "set wires" (*kuteya waya*) and snare small antelope, there was no relish—a health tragedy for people already in a serious famine. Bvekenya was not coming as a white man, but an adult man trying to hunt his way to fortune. Old Marhule treated him *semwana wemumusha* (like a child of the village), his wife cooked and served him food, and welcomed him from the forest with praise poetry just like she did her own son Mubhulachi, according to his own white ancestors. After all, everyone had ancestors. The meat he brought back home became the relish of the household; he was now the family's hunter.[13]

Marhule Mavasa may have been an old man but he had lost none of his cunning hunting skills. We saw in chapter 2 how maTshangana and vaShona hunters weaponized forest animals, principally the vulture, as a navigational aid to identify the site of a carcass in the forest. Old Mavasa was one of those old timers. Now advanced in age, he could no longer chase after animals, so he sought meat by other means. As Bvekenya was embarking on the hunt with Old Mavasa's son Mubhulachi, the old *sabhuku* beseeched him: "Bvekenya, please do not shoot my lion. I always take my share from its kills. It lives near here, in a certain place on the Lundi River which I shall show you." Bvekenya assured him that lion skins were of no value to him, but was curious about how such an old man—or any human being for that matter—could share in a lion's kills (Bulpin 1954, 124).

Old Mavasa volunteered no further information. Then, one night as the hunting pair roasted meat at a campsite, Bvekenya asked Mubhulachi what his father meant by "sharing" meat with a lion. Mubhulachi told him that on certain mornings his father armed himself with his bow and arrows and went to check if the lion had killed prey. Like all lions this one was also conservative; it hunted at specific waterholes, and never wandered far from its home range. All that Old Mavasa needed to do was to look up into the sky or the treetops for the circling or settled vultures. His wife hobbling behind him, the old man would approach the carcass, scaring the lion away with shouts and gestures with his bow and arrow. The lion would growl begrudgingly, then slowly withdraw. Whereupon Old Mavasa proceeded to "cut up the carcass, take what he considered to be the chief's share, give it to his wife to carry home in her tin can, and then leave the remains for the lion, who in a great rage had meanwhile watched proceedings from a suitable vantage point" (Bulpin 1954, 125).[14]

These encounters did not always end well, particularly when Old Mavasa found that the lion had killed a sable antelope on the Lundi River's banks, and dragged it away from sight, perhaps to hide it from the big cat's human nemesis. The gradient was steep, but the lion managed to secret its catch in a narrow, steep-sided donga and came out to the pool below to reward its thirsty throat. Old Mavasa followed the spoor. The lion heard him and beat a hasty return. This time the old man was not going to get away with theft. Meanwhile, Old Mavasa's wife Nwa Mukome arrived, saw the spoor of her husband—with the lion's right behind him. For anyone who knows such signs, they could only mean one thing: TROUBLE. Here was a situation—the carcass was located farthest away in the donga, then Old Mavasa, the lion, and then Nwa Mukome. As was custom, women were entrusted with managing fields; to scare away elephants and destructive quelea birds off the fields they beat empty tins with sticks, producing acoustic irritants that drove these problem animals away. That same tactic was used against lions threatening cattle.

On this occasion, Nwa Mukome beat her tin with the hunting knife she always carried when walking into the forest with her husband. She drove the lion toward her husband. Now *sabhuku* was cornered. The sight of his own wife, she whom he had paid bride wealth for, driving death toward him, was too much. He roared at her: "I walked to Kimberley in my young days. My brother and friend died of cold and hunger. I worked to get cloth and blankets with which to *lobolo* [marry] you, and now you drive the lion to eat me" (Bulpin 1954, 125). So ferocious was his ire toward his wife that it paralyzed the lion, which realized that it was caught between two human enemies, one with bow and arrow, the other a knife and lots of noise. Hunter became the hunted. Noticing the feline's quagmire, Nwa Mukome began to tactically withdraw, still shouting and beating her tin lest the lion think she was retreating in fear and charge. The lion followed, more to escape Old Mavasa through the entrance to the donga than to attack Nwa Mukome. Upon reaching the mouth of the donga, the old lady clambered up its sides to the top of the bank, then lowered a pole to rescue her husband—and the meat he carried (ibid., 124).

While Old Mavasa used animals as technologies of the hunt, Bvekenya tried to tame them into livestock. At that time there were teeming herds of *mhofu* (eland) on the other bank of the Muchingwidzi, right at Nyamare where even today children go to *guwa* (swing), and at Ngondomu, where a man named Gazini used to live.

Bvekenya had his *hachi* (horse, or mule), and he asked Old Mavasa for permission to capture and domesticate these *mhofu*. "How do you hope to catch them?" Old Mavasa asked. "I will chase them with my *hachi*," Bvekenya said. "Very well, if you can catch these animals, do it." So Bvekenya gathered his ropes, saddled his *hachi*, called Fifteen and Mubhulachi, and they went into the forest. When they arrived, they found the animals, and Bvekenya began chasing after them, on horseback. Cowboy-style he caught one *chimhofu* (young *mhofu*, a bullock) by the scruff of the neck. At that time Bvekenya used sledges for carting ivory to his hideouts. So they took a sledge and some oxen, yoked them up, including one on the same span with the eland bullock. When they brought the *chimhofu* home, they built a pen for it like that of the cattle. Mubhulachi and Fifteen cut green grass for the *chimhofu*, and gave it some water. It refused to eat at first, but hunger began to gnaw away at this despondency, and it began to eat and drink. Soon it got used to the idea. It began to become livestock.[15]

At one point Bvekenya increased his *mhofu* herd close to twenty through more captures. There came a time when Bvekenya decided his *mhofu* had now been domesticated enough and let them out to pasture. They did not run away, but grazed around the pens, and when the sun was setting they did not object to being driven back in. So Bvekenya employed Mubhulachi full-time as herdsman for his new livestock for an agreed wage, which he paid Mubhulachi's father without fail for years.

In time some *valungu* heard that there was a white man who was in good relations with *vanhu vatema* (black people) in *Rodhizha* (Rhodesia). *Mapurisa* (police) were sent down to check on this white man, to find out where he was coming from, what he was doing among black people, and to see if he was not a poaching suspect. So when these white men wearing *magondoro* (helmets) and *magomazi* (shiny brown shin pads) arrived and found it was indeed Bvekenya, a suspect in numerous poaching and public violence cases, they arrested him. Before he was led away, Bvekenya gave Mubhulachi a rifle as a parting gift. The young man said: "But what am I going to do with this gun? I do not even know how to use it." His father the chief Marhule said: "That does not matter my child; what matters is that you now have a gun. Give it to me and I will hide it—and then I will teach you how to use it." That is how Mubhulachi learned how to fire a gun from his father.[16]

Unlike Mubhulachi Mavasa, Watson Lamson Machiukele's father was already a distinguished marksman when Bvekenya asked him to accompany him on the hunt. Bvekenya employed the elder Machiukele to go on his own to hunt, kill, and bring the ivory, paying him a salary after it was sold. Initially, Machiukele had only hunted with Bvekenya because the white man had offered him a .303 rifle and ammunition and asked him to accompany him on the hunt. However, after the first kill Bvekenya talked Machiukele out of his plans to go to work in the mines of Johannesburg. Bvekenya offered Machiukele an escape from what was becoming a very difficult problem: finding markets for ivory as a black man. As a *mulungu*, Bvekenya offered a new possibility. The first two pairs of tusks sold at a decent price; with the money from his own pair, Machiukele managed to pay the *roora* (bride wealth) for his eldest son, Majuta. That was in the 1920s; by the late 1930s, Machiukele's younger son Watson would grow up

Figure 6.6
Watson Lamson Machiukele.
Source: BBP

herding his father's sixty-plus head of cattle bought with ivory money.[17] When Bvekenya left, Machiukele continued hunting, but more for meat than ivory given that the *mulungu* departed with his market, the store at Makuleke.

Ivory Processing and Trading

Bvekenya's priority was ivory. After the bull was shot, it was left to the African hunters in the employ of *valungu* to harvest the tusks and ready them for the market. One old timer describes the art of harvesting tusks out of a dead elephant: "After killing an elephant he used to leave it there to decay for a while without telling or showing anyone except us. Then in a few days we went in and pulled out the tusks from the loose putrid flesh. The hunter used to kill and dig a pit where he buried the tusks, covered the area with branches, and went away to continue the hunt. When the time for departure arrived he returned to the spot to take the horns."[18] The tusks were usually buried close to the path in bushy hides. Because their smell attracted scavengers and human attention, it was important to move them as soon as possible.

Ivory is very slippery and therefore presents many transportation problems. One method was to tie the tusks on the back of a donkey; however "their tapering smoothness always seemed to allow them to slip off" (Capstick 1988, 64). Another method was to drive wooden pegs into the tusks' open ends and attach ropes (made out of ilala palm leaves) at the thinner end of these tips. These ropes were then twisted and knotted at intervals around the tusks, before being suspended either side of a donkey. Heavier tusks were carted on a donkey-drawn sledge constructed out of the V-shape formed by a tree's stem and branches, finished off with two or three wooden crossbars. These sledges-in- motion furrowed paths along the forest that became permanent tracks and, eventually, roads (Bulpin 1954, 65).

After the elephant kill and harvesting of the carcass, the white hunter loaded both the ivory and the African hunters into his pickup truck. He would drive past the native commissioner (later district commissioner, the white colonial official in charge of "the natives," or Africans) or BSA police station and travel on the road freely without being arrested simply because he was white. "Even his firearm was not hidden. You would find the guns openly on the bed of the truck," recalls Rosa Khosa, an old hunter now residing in South Africa. "The white man was treated totally differently to us black people."[19]

From the 1920s right up to the end of Portuguese rule, the hunter-merchants and the storekeeper-buyers at Makuleke, Masenjeni, Mapai, and

Machaze sold their ivory at Lourenço Marques (and possibly other coastal towns like Vilankulos). The American safari hunter named Wally Johnson describes the details of an ivory transaction in the 1930s:

The local Indian ivory dealers knew they had a captive market and they exploited it. The going rate was one-pound sterling per pound, but not in L.M. [Lourenço Marques]. There, three Indian ivory dealers were cahoots and worked together. . . . You'd go to the first one, who would offer one shilling and three pence a pound. He would say, "Mr. Johnson, you've been in the bush so long, hunting elephant, that you haven't heard the news. The price of ivory has dropped drastically. I lost money on my last transaction, so you'd better take the tusks to somebody else."

The next trader would say, "Oh, I've lost so much on ivory, you must take it to somebody else."

The third one would offer a couple of pennies more, and you were forced to take it, as [ivory] was so weighty and expensive to transport. They were smart bastards, all in telephonic communication with one another as I went from one dealer to the next; they, in the meantime, figuring, quite literally, what to offer me. They got stinking rich. After ten months' hunting, I started to believe that the price had indeed plummeted (Capstick 1988, 81–2).

The Indian diaspora in southern and eastern Africa is famed for commercial enterprise and the ability at bargaining with customers (Kosmin 1975). From the late nineteenth century onwards, they came to dominate the wholesale sector, owning entire streets and sectors of both urban and rural southern and eastern Africa by means of buying out their competitors and networking the shops into a cartel capable of outstocking and outpricing any remaining shop owners. Even today, when a customer in one shop does not find the right size of garment or shoes, she is told to wait while the shop owner sends an African assistant to collect it from the next shop or, as in this case, uses telecommunications to instruct that the item be brought over. The Indians are the *buya tinapangana*, Kiswahili for "come, we'll talk," a tribute to their powers of bargaining.

From the 1930s through the 1960s, the Indian shop owners used more than just persuasive skills. If the ivory hunters tried to turn violent, these Indian dealers had the protection of African gangs with rifles (Capstick 1988, 82).

White Poaching and the Game Reserve Idea

The earliest attempts to establish a national park in Gonarezhou were anti-poaching moves against white rather than black poachers. The establishment of Kruger National Park in 1926 had a knock-on effect on the three neighboring states' border enforcement. With the introduction of game laws on the south bank, Gonarezhou became the nearest place for white South African men craving "a shot." Kruger National Park had inspired the Wildlife Protection Society's Southern Rhodesia chapter and the Commerce Department into exploring the modalities of establishing a game reserve with the picturesque Chipinda Pools along the Lundi as a nucleus. Backed by individual hoteliers and safari operators, the Department of Commerce was specifically interested in making Chipinda Pools the center of a game reserve. In June 1928, the chief entomologist recommended an even bigger sanctuary on condition that contiguous land "not good for anything else" was identified.

Early calls for the establishment of a game reserve in Gonarezhou resulted overwhelmingly from growing concerns about poaching. In 1934, some seven thousand Africans were living in Gonarezhou. Chief Chitsa's people at the Save-Lundi junction were "experts with the bow and arrow and the setting of traps"; they were killing "large quantities" of animals. To the south of the Lundi was a compound "where natives collect[ed] for transportation by motor lorries to the Limpopo and Parfurie [sic, Pafuri]"; these people, in particular Ngwenyenye's, hunted "entirely uncontrolled" and showed "no respect for a white man" (Bulpin 1954, 151). Another culprit was Matombo and his people on the Mwenezi, who killed "at least three big buck per week" with snares, traps, dogs, and guns.[20]

While the Department of Agriculture and Lands and the Department of Commerce were calling for these villagers to be removed, the Native Department insisted on the allocation of alternative land to settle them first. It would not promote overcrowding, soil erosion, and overgrazing in the Native Reserves just to save "wildlife." The department of Agriculture and that of Lands and Commerce argued that the envisaged 50- to 100-mile fencing would keep lions, hyenas, wild dogs, and leopards away from these villagers' 3,000 head of cattle. Native Department disagreed: the fence would render untenable the existence of 1,500 Africans within Crown Land now proclaimed for the game reserve. The department would be more than happy if these people were removed because it would not have to worry about policing these remote, unreachable localities. However, the whole plan made no sense unless the Department of Agriculture and Lands supplied the land first.[21]

The proponents of the game reserve scheme needed a more powerful argument than just conserving nature for its own sake. Two developments took place beginning in 1929 that changed the situation in favor of transforming Gonarezhou from "unalienated land" to "game reserve." First, the

Great Depression struck the world markets, severely reducing the demand for agricultural products like tobacco and beef. The Rhodesian government scampered for alternatives. Second, in 1930, Rhodesia introduced the Land Apportionment Act, which re-designated Gonarezhou from "unalienated" land to "'unassigned area,' . . . neither African, European Area nor Forest Area." The move left Gonarezhou open for alternative land uses like conservation and tourism ("Rearrangement of Portfolios" 1933, 432).[22] Still, tourism in Gonarezhou would generate a fraction of the amount needed to combat the losses to agriculture.

New plans emerged to link the project to a bigger transfrontier conservancy in 1934. The project was essentially a South African attempt to extend Kruger National Park into neighboring areas to absorb its overstocked "game" in that park. The Department of Agriculture and Lands hoped to cash in on Kruger's expansion and tourism business by offering the two million acres of land between Save and Limpopo, between the border and the western fringes of the Mwenezi. The proposed Gonarezhou game reserve was now composed of the triangle of land between the Limpopo, Mwenezi, and Save rivers touching the Kruger (Kruger National Park 1934, 299).[23] The objective was to create a continuous transfrontier game corridor from Kruger right up to the east coast of Tanganyika ("Publicity Collaboration" 1937), with Gonarezhou being developed for tourism.

Borrowing from the British East African model, Chipinda Pools would act as the core of a game reserve and safari area where tourists would come to watch or hunt animals, to sleep, eat, and enjoy their holidays.[24] Two structures were recommended for construction: a "country hotel . . . with mosquito-proofed accommodation" built in the beautiful chevron style of Great Zimbabwe monument, and a store to provide supplies for tourists.[25] Proposals were put in place to station a warden with police experience at Chipinda given "the often lawless nature of this border area."[26] "Reliable natives" would look after visitors, giving them confidence to camp and explore the reserve.[27] There was one problem: the Department of Agriculture and Lands could not agree to subsidize the incentives the Department of Commerce was asking for. It was the view of veterinarians that the game reserve would become a reservoir of predators and veterinary diseases (principally foot and mouth disease [FMD] and potentially trypanosomiasis or *nagana*).[28]

This is exactly what happened. By 1937, the game reserve proposal had been completely abandoned because of veterinary disease implosion despite the persuasive arguments of the Department of Commerce and the Southern Rhodesian Wildlife Protection Society (Kinnell 1958, 155–157;

"New Zambezi Bridge" 1934, 103). Nuanetsi Ranch, the country's premier beef producer, had declared losses of six hundred head of cattle per annum to carnivores and tick-borne diseases spread from forest animals. It had struggled to cope with FMD attacks since 1929; the 1932 strain had also infected 800,000 head. After a respite in 1933, the disease broke out again in 1934 ("Stock Restrictions 1933, 22).[29]

Tsetse fly sealed the fate of the project. In 1935, the Department of Tsetse Control commenced operations to clear forest in the Chipinge District to check the tsetse fly incursion from Espungabera (formerly Chipungumbira). Trypanosomiasis was claiming "many cattle" each year, and the chief entomologist warned of "a menace not only to Rhodesia but possibly to the Union as well." He did not wish to create a highway for tsetse into South Africa ("Development of Game Reserves" 1937). The proposal was dead in the water. For the next two decades the Rhodesian government battled tsetse fly, any game reserve being completely out of the question.

Conclusion

This then is how indigenous hunting was systematically suppressed as criminal acts that were, as Wright put it, informed by "absolute fatalism," "pure cruelty," and "without feeling whatever for animals." Through their actions and at their *huvo* and other discourse spaces, maTshangana made the counter-proposition that the criminalization of the hunt was not only unjust, but also ran against the grain of history. The people were living or hunting inside the forest before the *mulungu* arrived, and they found a surfeit of "game" in all varieties. If African hunting was such a danger to forests and animals within them, how is it that Europeans found these animals in such abundance? What factors had changed that made African hunting so destructive—or was "destruction" a way of seeing, a difference of philosophical approaches to the forest, regarding who ought to move within it, and what constituted proper movement in a space like that, as well as who owns this space and by what right?

To these maTshangana and vaShona, the recreational and the aesthetic were not the only purposes and meanings of the forest, which was also a spiritual, extractive, and educational space. To the *mulungu*, nature was the recreational, the picturesque, and the safari and African's spiritual beliefs about and practices in the forests were mere "myths." Fences had to be erected to demarcate the jurisdiction of nature, on one hand, and the beginning of the "tribe," on the other. Mixing nature with "these tribes" bred chaos.

But whose culture was nature? Certainly white culture, as Jane Carruthers and Terence Ranger have shown in the cases of Kruger and Matopos National Parks, respectively (Carruthers 1995; Ranger 1999). According to District Commissioner Allan Wright (1972, 142–143), Africans had no sense of nature, the aesthetic, and the recreational; when they saw an animal, all they thought about was the cooking pot, and they quickly rushed to grab their snares. The white man had come just in time; had colonialism come a shade later, at the rate at which Africans were acquiring guns and wires from the mines of South Africa, the odds were that "African wildlife" would be staring into extinction's menacing eye.

Poaching as critique took on a different tone. To begin with, it was precisely this profane view of the forest—as simply a collection of beautiful trees nicely juxtapositioned according to hills, and offering such "fascinating" and breathtaking beauty to the eye—that was the problem. It meant that when that same forest and its animals became contaminated or imperatives of making wealth necessitated it, the aesthetic could be sacrificed. That is what had happened in the 1930s through the 1960s. In the 1920s and 1930s, Africans watched with mouths agape as truck after truck of South African *mabhunu* careened past their villages filled with dangling carcasses. They protested at the white man's wastefulness—of slaughtering a whole elephant and leaving its giant body to rot so long as they got its two small "teeth." When it was convenient, the white man not only authorized Africans to kill all kinds of animals, but also armed and supplied them with "modern" guns and ammunition, because he had run out of ideas about how to deal with an insect called tsetse fly. Was it not the same colonizer who had defied the wisdom of the ages about the sacredness of certain forests as reservoirs of bad spirits and illnesses that must be left alone, and gone ahead to partition them into farms and international borders?

The idea that Africans had no sense of demarcation between the village and the forest, that they lived in the bush like animals, and that they were part of the picturesque—what Wright called "wild Africans"—ran contrary to what we saw in chapters 1–3. Those procedures of approaching and moving within the forests, dissolving and respecting human-animal or village-forest boundaries, occurred under the guidance of *Xikwembu* or *Mwari*. The white man left no room for the ancestral spirits and God in the national park; the hunter approaching the park saw himself as returning to and walking within the sacred home of his ancestors. After all they were all buried here—graves overgrown with weeds and brush, but spirits ever alive, guiding and looking out for the hunter, pointing him in the direction of his fortune, which *valungu* interpreted as poaching.

To the hunter, nature was a totem, food, clothing skins (if only now ceremonial), and an educational space where boys were groomed into *amadoda sibili* (real men). The pot was not just the *raison d'être* for hunting, but a sign that the education handed down across generations, sometimes from father to son, other times through communal knowledge, had achieved uptake. There were rules governing the hunt: breeding seasons were known, closed seasons respected, and open seasons fully utilized. People hunted just enough for their needs, and the animal population was too big and diverse to be exhausted.

Then the white man came with his gun, his rinderpest, his lorries, his monopolistic laws, and his tsetse operations. All hell broke loose. The animal populations diminished. The very same people who had caused the massive depletion of herds now became the fiercest warriors for conservation. The very same people who had acquired their skills and knowledge of Africa's forest animals from Africans now criminalized the very professoriate of the hunt that had educated them. They had found a society where animals, like all natural resources, was like water—so critical to survival that without it people must surely die. Now they were telling people not to drink water.

7 *Chimurenga*: The Transient Workspace of Self-Liberation

Poaching is always political. It is not merely about craving for meat that leads to tracking, snaring, and killing without seeking permission from the law. To the local inhabitants, Gonarezhou National Park was ancestral lands from which they had been forcibly evicted at gunpoint. Their shrines and ancestral burial grounds, the venue of the hunt, and multiple forest resources they had relied upon were now on the other side of the fence. In circumstances like these, poaching became political critique through defiant action. The forest was a sacred space; it belonged and would always belong to the ancestors, fence or no fence, *mulungu* or no *mulungu*.

MaTshangana and vaShona around Gonarezhou had felt this way since before the declaration of the proposed game reserve, during the time when District Commissioner Allan Wright was ruthlessly disarming them of their guns. It was then that an important event happened: the national ferment of resistance against colonial rule merged with the local grievances, turning poaching into a transient workplace of anticolonial resistance. Since the 1930s, the emerging African elite educated in universities in South Africa and America had used the "civilized" language of diplomacy without success. Tired of talking, the African nationalists began to mobilize for civil disobedience and military action if necessary.

Fortress Conservation and "Poachers"

Between 1937 and 1967, the Chipinda Pools area—indeed the Gonarezhou project—was mired in anti-tsetse operations discussed in chapters 5 and 6. Only after extensive pesticide spraying, vegetation clearance, and massive animal slaughter was it possible to realize it. In 1968, the Gonarezhou Game Reserve was officially proclaimed. It became a national park in 1975.

Until Wright arrived in 1958, a free-for-all hunting bonanza was the order of the day in Gonarezhou. While the colonial authorities were preoccupied with tsetse fly, hunters of all kinds descended upon the areas south of the Lundi River to engage in what the state called "poaching" (unlicensed hunting) but which the participants saw as simply hunting. The state had no mechanism to control the hunting, thereby rendering academic the (il)legality of the hunt. Moreover, government officials regularly brought their white visitors to enjoy a "bushveld life," away from the madness of an industrializing Salisbury. Scientific expeditions involving schools and national museums also had the liberty to shoot for laboratory specimens (Strover 1958, 32). Local white ranchers proudly kept displays of lion skins and elephant tusks in their houses (Cary 1954).

It is not for nothing that the local vaShona and maTshangana gave Wright the nickname *Chibwechitedza* (Slippery Stone). To some locals, he was and is still thought of as Rhodesia's president.[1] He is remembered as a very cruel, hard-hearted *mulungu* who as a magistrate was indeed a slippery stone: when facing him, it was impossible for an African accused of a crime to convince him of his innocence. He is most remembered for caring more for forest animals than the Africans he was being sent to administer under British indirect rule. It was in his administration that the criminalization of the professoriate of the hunt as poaching really gathered pace. Wright basically divided *mapocha* into two: the local villagers and whites (Simpson and Booysens 1958, 58).[2] Using roads, radios, transport, and bicycles, and six World War II surplus "No. 19" VHF/HF radio sets imported from Cape Town, each with a 400-mile reception range, he connected the various isolated suboffices into one administrative circuit, with his station at Nuanetsi as a command center (Wright 1972, 210).

As interpreted by locals subject to Chibwechitedza's anti-poaching network, the game reserve was not established to preserve animals but "*kuti abate vanhu vanovhima*" (to catch blacks who hunt). The reason why *valungu* found the animals was because they were already being preserved.[3]

Throughout the 1960s, the Department of National Parks and Wild Life Management (DNPWLM) responded to distress calls from villagers losing cattle to predators and crops to elephants and buffalo. Scouts now had their hands full trying to control "problem animals" straying and causing mayhem in the villages.[4] In 1969, one white research officer had to be airlifted by helicopter to hospital after an elephant first gored then "attempted to kneel on him, fracturing his pelvis."[5] In the same year, staff members at Mabalauta were preoccupied with calls from villagers and ranchers seeking to drive off or destroy elephant damaging fences and irrigation installations.[6]

While kicking everybody else out, there were some Africans that Wright tried to keep inside Gonarezhou. These "wild Africans" would add to the picturesque "nature" that was emerging, and he wanted to preserve them in their natural state, alongside other "wild animals" like lions and impala. To him, maTshangana of Ngwenyenye were still "primitive, ultra-conservative, unspoiled Shanganes (sic, maTshangana) living as they had [been] a hundred years ago" and were "part and parcel of any national park scheme of the future" (Wright 1972, 65). The overseas tourists, who had a surfeit of dams, towns, buildings, and mountains at home, would seize the opportunity "to study wild animals and 'wild' Africans." They offered Gonarezhou "a wonderful opportunity to combine the two great attractions in a unique and beautiful setting." Moreover, these people were just another species of predator, whose poaching was part of the ecosystem's checks and balances. DNPWLM saw some sense there: their poaching activities would save the department "the trouble of having to cull later when the area becomes over-stocked" (ibid.).

The racism in Wright was felt accordingly in the ways he treated his African subjects. It exemplifies what I have described elsewhere as the descent of humans into human game, to be preserved as an attraction and treated as vermin to be quarantined or destroyed when deemed pestiferous to the state (Mavhunga 2011a). This "noble savage" tradition, in which the colonizers froze Africans living in animal-rich areas in time and extolled their captivity to and "harmony with nature," was marked by a contradiction (Brown 1983; Voss 1982). Namely, that the colonizer conveniently isolated those aspects of "the native" that suited the project of domination, but cracked down on those deemed detrimental to the colonial project. Taken to its logical conclusion, the language of "wild Africans" was not only racist and paternalistic, but it also designated Africans as nonhuman aesthetical objects whose life could be spared or taken without consequence (Agamben 1998, 2005).

Locals interviewed in the Chibwedziva area of Chipinda Pools recall the unprecedented war Chibwechitedza instituted to disarm them of their guns, snares, and land in order to "save wildlife" from "poaching." Many maTshangana who owned guns never disarmed in the 1890s, but simply buried them in fear of Chimamise, Wright's African head "district messenger," the title given to the pseudo-police force composed of local African men.[7] Chimamise means the one who induces such fear in people that they soil their pants at the mere sight of him. Wright was purging what he saw as an unholy alliance between "tradition" (knowledge of the hunt) and "civilization" (guns, cable snares); to save animals, that link had to be broken. One anthropologist had observed in 1954 that while the local inhabitants had generally retained "much that is picturesque and primitive," they were "rapidly being 'civilized.'" Their "tribal society" was "in an advanced state of disintegration. It would not be long before their way of life [was] completely swamped with the new set of values which they must absorb from mine, farm and native store" (Philpot 1954, 47). The absorption of such values and the choice of what to absorb had to be controlled, otherwise the good "native traditions" would be lost along with the bad.

Wright's "wild Africans" idea was incompatible with the DNPWLM's view that "game and people cannot live satisfactorily side-by-side" (Wright 1972, 65) and the Department of Agriculture's concerns regarding the veterinary threats Sabhuku Ngwenyenye's cattle posed. The community's protests that their "great grandfathers had been born there, lived there and were buried there" fell on deaf ears.[8] DNPWLM maintained that there was no need for human settlement to impinge on the unbounded mobilities of forest animals, hence at Marumbini "where elephant herds were previously unknown, large herds were occupying the area only eight days after the squatters had moved out."[9] This was of course not true; elephants had always roamed the Marumbini area.

Inevitably, the arbitrary manner in which the fences around Gonarezhou were established—against the wishes of Africans, and with the intention to deny them access to forest animals—set the park up for transgression. The escalation of poaching was a reflection of resistance against the forced removals from the area that was designated Gonarezhou Game Reserve in 1968. The following year, sixty-three Africans were arrested in the game reserve and adjacent areas for contravening the Wild Life Conservation, Fish Conservation, and Forestry Acts. However, the main headline was a serious case of assault on a game scout who was trying to re-arrest one man for poaching: "he suffered five axe wounds in the head, had his clothes removed and was left for dead." Regional Warden Douglas Newmarch noted in 1970 that "snaring and poaching" in Malipati was out of control because the Ngwenyenye people resettled there were "an embittered lot" (Wright 1972, 23). This general feeling of disenfranchisement was not only material (the denial of hunting) but deeply spiritual and political: by eviscerating maTshangana and vaShona from where their ancestors were buried inside Gonarezhou, *valungu* struck at the heart of their very being, of who they were as a people.

Chibwechitedza wemaTshangana

This time something had changed. In times past, maTshangana would have been content to disobey *valungu*'s law and continue hunting. They would have allied with other *valungu* like Bvekenya, whom they would make *dhulu* (granary) out of. But now Bvekenya was gone, there were no animals left in the adjacent Portuguese territory for these *mapocha* to hunt. Bvekenya himself had just died in 1962, having long retired to the life of a farmer in the Northwest Province of South Africa. The war of independence had just begun in Mozambique, but far away in the Zambezi valley, too far to even induce any echoes locally.

As we saw in chapter 3, since the 1700s maTshangana (then known as maTsonga, and later maHlengwe) had developed a habit of tapping into outside (incoming) actors to derive more value from their forest resources than they otherwise would on their own. They had done this with Dutch traders, then with *maputukezi*, and more recently Bvekenya. These expressions of initiative, whether as vassals of the Gaza or lately the BSA Company, point to a tradition of designers using outsiders as sources of *nzira* (means), if not the *nzira* (path) itself, to prosperous futures. This emerging tradition of extraversion was of course mutual; the outsiders coming in were also using these locals as means and paths to their own ends (Mavhunga 2011b). Halfway into his term as district commissioner, Allan Wright was about to deal with the latest incarnation of this tradition.

It started far away from Gonarezhou, in the cities of Salisbury, Bulawayo, and Umtali. Invoking the emergency provisions of the Law and Order Maintenance Act (LOMA), Prime Minister Ian Douglas Smith had ordered that the entire top echelon of the African nationalist movement be arrested and squirreled away to very remote places, far from the cities where they were inciting urban Africans to violence and acts of civil disobedience. This followed an escalation of clashes involving the two major nationalist parties, ZAPU (Zimbabwe African People's Union) and ZANU (Zimbabwe African National Union). Having peacefully agitated for independence, these nationalists had become alarmed that instead of granting independence, the white minority regime was instead moving toward declaring independence from Britain and remaining a white governed republic. The gloves came off.

Joshua Nkomo, leader of the now banned ZAPU, was one of those arrested in a pre-dawn raid on April 16, 1964, along with his deputy Josiah Chinamano, his wife Ruth, and countless others. They were flown to Gonakudzingwa, one of the several designated "restriction camps" that also included Sengwa, Marandellas, Nkai, Connemara, Wha-Wha, Chikurubi,

Figure 7.1
Joshua Nkomo (left) dressed in animal skin hat in the 1960s.
Source: Zimbabwe Review (January–February 1970)

and Gokwe, to "cool off" (Nkomo 1984). Here, African nationalists would be detained without trial in huts less than a mile from Malvernia. To the west of Gonakudzingwa, abutting Gonarezhou's Chipinda Pools section, lay Chibwedziva area. Remote places like these, surrounded by elephant, lions, wild dogs, and other beasts of the forest, were deemed the perfect places to send these hotheads. The problem was this: the court order allowed these men freedom to continue their political rallies and organizational work tens of miles from the restriction camp.

A feisty politician carrying incendiary ideas was thrust in the middle of an ongoing struggle between Chibwechitedza and local maTshangana over Gonarezhou's forest resources. This then was a period in which indigenous institutions were extended to serve as spaces for purveying, hosting, and executing resistance against the colonial settler state. There were basically two major phases in this resistance. The first was a period of mutual assimilation between national and local struggles, with spaces that we have seen before—*nzira, huvo, dhema,* and the hunt—being extended to seek, digest, and respond to what the detainees at Gonakudzingwa were saying. The second phase from 1976 until independence in 1980 involved the radicalization of resistance into open guerrilla warfare, in which the endogenous institutions now anchored the war itself. If chapter 6 sought to appreciate poaching as critique, this chapter seeks to place the outcome of that critique—resistance—within the context of other rebellious mobilities of their time. The starting point of that long discussion, however, is more mundane.

When the men of Masivamele (Chibwedziva area) heard of Nkomo's presence, they rode to Gonakudzingwa on their *xikanyakanya* (bicycles, especially of the 28 Humber type) to hear him speak. From the paths leading out of their own homesteads and villages they got into the Chikombedzi, Chibwedziva, and Machindu roads, the very same that Ndambakuwa, Chibwechitedza and Native Commissioners had forcibly driven locals like slaves to build.[10] In the morning the men took off, riding their bicycles in single file—the very same bicycles they used to transport carcasses and sacks of meat they killed in the forests of Gonarezhou. On their person they also carried *busva* (or *sadza*, the traditional meal from ground corn, millet, or sorghum) to eat along the way, the relish being salted and smoke-dried meat—of buffalo, antelope, or buck. The African footpath (*nzira*) across the forests was one single thin furrow less than eleven inches thick, indeed a self-made pathway to liberation.

Food would be tied onto the bicycle carrier, some *mangayi* (boiled maize) here, *svinyama* there, *mati* (water), and other eats, over there. Along the way, they stopped in the burning morning sun, wiped off the sweat from their faces with their arms, and sat down to eat, before starting off again. In their pockets they also each carried a financial donation to the party. As they rode excitedly to Gonakudzingwa, the men of the village would be "talking politics." A primary topic as they crossed the Gonarezhou was how Nkomo would help them recover their land, restore dignity to the graves of their ancestors now buried in bush, and to themselves, and the spiritual and environmental practices that defined them as a people.[11] It would be a *quid pro quo*: they would support Nkomo *if* he promised to help them get back their ancestral land. He promised.

Here was "*Chibwechitedza wemaTshangana.*" As vaShona would say, *goné ana goné wakewo* (even the most powerful man has his own match); Chibwechitedza had met his. The poor rains and complete crop failure of 1964 made Nkomo's message of national liberation resonate even more among the locals. The three chiefs in Matibi II, Chilonga, Mpapa, and Masivamele, had requested government aid, and Chibwechitedza had failed to deliver (Wright 1972, 368). Nkomo beseeched the locals to engage in acts of civil disobedience—to hunt as much as and wherever they pleased, refuse to dip their cattle, and to pay taxes (ibid., 372–373).

The locals enjoyed seeing "Nkomo," clad in his skin hats—products of their own professoriate of the hunt. In the 1960s, educated African nationalists had returned to a more assertive identity celebrating their *ntumbuluko* (culture). Many African elites had initially resorted to aping whites hoping that "attaining a certain level of education and finance might be seen as a rite of purification (white = pure, black = polluted)" (Fanon 1952/1967). By 1964 the conformist discourse had given way to a return to an *authentic* African identity. The animal skin had become a trademark of "the African nationalist," and there was a brisk trade locally in spotted animals like genets, leopards, and cheetah. Indeed, Nkomo came to address gatherings at places like Chikombedzi dressed in his own skin hat (see Figure 7.1).

The Return to the Professoriate: Localizing Nationalism and Nationalizing Localism

In truth, *ntumbuluko* had never left maTshangana but had always acted as an institution they constantly remodeled to "anticipate a future saturated with projects of an indisputable modernity" (Diouf 2000, 683–684). Being at Gonakudzingwa and campaigning in places like Chikombedzi, Nkomo and his political elites could negotiate a return to traditions that they had earlier looked down upon as unsophisticated and feudal in preference to the modernity of Western learnedness.

History so thoroughly professoriate, so immersed in ancestral idioms, had not just "stepped aside" for the colonizer who was, after all, seldom present except as a pest in people's lives. Nkomo did not set aside that history of guided mobility when bringing in his "nationalist" message; he clothed nationalism in chiTshangana (Tshangana idioms). He would appear to maTshangana not just as a human being but a human being with very special powers, powers more powerful that those of *valungu*'s god. Chibwechitedza did not work visible miracles; people say Nkomo did. It was part of the *masaramusi* that he staged to convince people of his supernatural powers and his ability to free black people from oppression by *mabhunu*. Nkomo called himself the *Mambo*, title of the king of the powerful precolonial Rozvi Empire of vaShona that once covered an area bigger than *Zimbabwe*, the name the nationalists were calling Rhodesia. Zimbabwe itself was an African colossus; it was the precursor to the Rozvi Empire.

Nkomo was the *Mambo*-incarnate, and he was coming "to take the land" (*wasviteka tiko lahina*).

Nkomo looked the part. In his hand the *Rozvi Mambo* also held the *ndonga* (*tsvimbo*, or walking stick), on his head rested snugly a leopard skin, given to him by local chiefs coming to Gonakudzingwa to pay their respects to their *Mambo*. In the meetings, locals say Nkomo would take off his jacket and hang it in the air.[12] And there it hung until the meeting was over. Then the Rozvi Mambo took it, put it on, and bid the crowd goodbye with hearty exhortations of *"Zii, Tiko Lahina"* (Zimbabwe, this country is ours). The people were told and believed that Nkomo *"utasviteka tiko lahina"* (he will take this country), because *"haapfuriki nepfuti"* (he is impervious to bullets). Nkomo was the invincible man. He was indefatigable: those that did not want him in a meeting would just find him there. They could not see how he would have entered. Nkomo would just blind them and get in, and with that demonstration, he gave them a hint of the untouchable fighter for African freedom. That power would rub onto anyone who undertook the pilgrimage to Gonakudzingwa to see him, purchased party cards, and carried the burden of his message around (also Wright 1972, 368).[13]

The use of magic or supernatural occurrences is something that we have already seen maTshangana and vaShona associating with the forest as a sacred space back in chapter 2. Talking to the elderly who carry forward the professoriate of indigenous knowledge in 2011, it was very clear that the practice of magic or supernaturally guided feats cannot be separated from the foundations of maTshangana and vaShona societies. Stories are told of people that cause their foes to see flooded rivers in the dry, rainless winter that force them to return home, even as others are walking across the dry sandy riverbank; of a man from Maxaquene, Mozambique, who lies down, is set on fire, but who rises, dusts himself off, and walks away; who uses hot glowing embers to scratch his itchy skin; who holds off the bus from moving by just standing on the road; who lies across the road and lets a bus or truck run over him, then stands up, dusts himself off, and walks away; who single-handedly carries on his shoulders a heavy sack that three strong men would struggle to lift—people who could turn sticks into snakes. All this, just to show one possesses supernatural powers and that adversaries must back off.[14]

As a *nzira* of political mobilization, Nkomo's magic lay in swearing people to secrecy, but there are fewer people able to keep secrets than maTshangana anyway. Chibwechitedza had formed a comprehensive and active intelligence organization comprising teachers, dip tank attendants, storekeepers, government agricultural employees, *masabhuku* (village heads), and *hosi* or *madzishe* (chiefs). His *zvikonzi* (Native or district messengers) were the runners moving between these surveillance assets and reporting to Chibwechitedza via his feared head messenger, Chimamise. But Africans told him what he wanted to hear; he failed to get through "to real African thoughts and feeling" (Wright 1972, 215–6). Whenever people were asked, no matter the intimidation or violence, the answer was standard: "*Anisvitivi*" (I don't know).[15]

MaTshangana's network of information security and dissemination was deeply grounded in the key educational sites we saw in the first few chapters. They had never died. At every *huvo* in Masivamele, the talk was that Nkomo was at Gonakudzingwa, "he is doing politics, he is fighting white oppressors so that black people will be free, so we should go and get his cards."[16] The cards Nkomo was telling people to come and get were pictures taken of him wearing a skin hat, a leopard skin draped all around him, products of the professoriate of the hunt. They were not cards as such; just photos. Each person had to get one in order to confirm themselves as supporters of the struggle and to receive *shavé* (spirits giving the receiver special powers) to resist *mabhunu* and so that the ancestors could guide them in their mobilities in service to the liberation struggle.[17]

Titus Mukungulushi of Chikombedzi was a regular visitor to Gonakudzingwa and brought back messages for Sabhuku Masivamele in secrecy, for the *zvikonzi* would be listening out through informers. Chibwechitedza would then send his *zvikonzi* to arrest the plotters. Mukungulushi was the one who initially attended Nkomo's sessions at Gonakudzingwa and then instructed people to go as well and see for themselves. "*Mukungulushi wauya*" (Mukungulushi has come) became the code for men of the village to prepare for the trip to Gonakudzingwa.[18]

From the *huvo/dare* of Sabhuku Masivamele, the trusted ones would be informed and brothers and friends would visit as if for a sundowner at the *dare*. When they came to the topic of Nkomo, the men would descend into low tones lest someone overheard them, or their sons sitting with them spoke about these conversations when with peers whose parents were "sellouts."[19] When someone had a visitor at his *huvo*, fathers or sons would arrive to get party cards. The men would discuss Nkomo's *mashiripiti akawanda* (many magical acts). The *zvikonzi* of Chibwechitedza knew that ZAPU cards were being distributed, and they would mount inspections, turning every house upside down. People ended up hiding the cards outside their houses.[20]

Humwe or *nhimbe* now became a space for political discourse. The topic of "Nkomo" animated conversations while *Zii tiko lahina* became part of the songs sung in this important, age-old transient workspace, and other adult gatherings that involved beer but whose purposes were not work-related. Hence Nkomo was also a subject at *padoro rematanda* (beer of the logs), the traditional beer brewed from millet and *rapoko* that people drank at the host's homestead while sitting on logs.

Since time immemorial, women were the *nyanzvi* at brewing *nkanyi* or *marula* beer and *mlala* palm wine (Bulpin 1954, 22). The *marula* fruit ripened in January and February into a tantalizing orange color, and the locals gathered it in baskets, cooked and sieved the residue into huge pots, and left it to ferment into sweet wine. They also drained the sap of *mahanga* (*ilala* palm) to make *njemani* or *chemwa*, another highly intoxicating wine. When someone wanted a field plowed, weeded, or harvested communally, they made a beer call (*kusheedzera doro*). At the crack of dawn everyone would be assembled, their oxen ready to plow, *mapadza* (hoes) ready to weed, *matemo* (axes) ready to chop. The plowing, harvesting, cutting of forest to make fields, home construction, and thrashing of grain all finished, the beer and food would come out, and the adults would drink like babel fish. The youths were given *bumhe* and *mahewu*, two nonalcoholic brews made from rapoko and millet malt.[21] There was no bottle store for selling industrially produced liquor because Africans were not allowed under law to drink lagers, wine, or spirits, so every beer call, for *humwe* or for sale, was always well attended. Every woman was a brewer, or else she was an embarrassment to her people. What could be worse than witchcraft! In the 1960s, some sold such beer, one pint for sixpence.[22] As people drank, they were talking politics.

Soon Nkomo had also become part of the games children had long played. MaTshangana children had long played *nkhati* or *dhema*.[23] Timothy Sumbani of Chipinda Pools is one of five sons of Dick Sumbani who grew up playing *dhema* in his home area on the Muchingwidzi River. *Dhema* was played during *tirhisa tihomu* (herding cattle), on the way home after school, during weekends, on school holidays, and even during "break time" or "lunch-time" at school.[24]

The boys would uproot the *nkhati* to reveal a tennis ball-sized bulb. Each boy would arm himself with sticks strong enough to hit the *dhema*. Next the teams were chosen, some this side, others that side. The game of the forebears would begin in earnest. One player on Team A would throw the ball to the right hand of a Team B player, who must hit it. Miss and it's a score, and it was the Team B player's turn to throw; Team A had to defend themselves. If the defending (Team B) player hit the *dhema* with the stick, he immediately left the *dhema* there and took up his position on the other side and waited with his stick for the Team B player to throw the ball. *Dhema* proved difficult to get, so the boys "started making a round wheel with grass, and then roll it with strings until it is strong, then we throw it." The opposing player still hit it like *dhema*, and ran and took position to receive the throw in turn.[25] It came as no surprise that Nkomo and his nemesis Chibwechitedza became readily available as names for the opposing teams in *dhema* and other games.[26]

Nkomo and his namesake Chibwechitedza also found themselves on opposing sides, of a game played with dry cow dung. The teams chose their sites well—especially by the seasonal pan, where cattle converged to drink water, or pools along the Muchingwidzi and Lundi, where huge amounts of dung accumulated. Each side stockpiled its "ammunition" (cow dung), and then the battle began. When one side won, the players changed sides, as they continued skirmishing and exchanging "fire" down the *vlei*.[27] The teams would continue their competition on the shimmering pools of the Lundi, "Nkomo" team versus "Chibwechitedza" team throwing rocks at schools of hippopotami and swimming in the big river's tributaries. Those that left the river or pan as "Nkomo" arrived in the village as "Chibwechitedza," and vice versa.[28] Some "games" often ended up inflicting a trifle more pain than necessary, and fistfights occurred that were, after all, part and parcel of being (boys) and becoming (men). Seldom did they become permanent enmities; that is what *vakoma* (elder brothers) ensured would never happen (Chipamaunga 1983, 11).

While the boys chased each other across the forests and prairies, the girls would usually be busy helping their mothers or playing their own games. They went to pick up and harvest wild vegetables like *nyapape*, a climber; *chibhave*, which is sour; and *dhelele / derere* (wild okra) leaves. They were also the pounders of grain; afterward they took clay pots and headed for the wells to fetch water, which they carried back, balanced on their heads without any hand support, and, upon reaching home, swept the floors and cooked meals for everybody. In their spare time, usually in the afternoon they went to the Lundi and Muchingwidzi to do the family laundry in the river. In the hot sun, with all clothes hung onto the branches and shrubs to dry, the girls dove into the pools reserved only for girls and young women. No boys were allowed within any visual radius. Here the girls also sang the new African revolutionary anthem: *"Tsuro-tsuro we-e tatore nyika, Tsuro-tsuro we-e naNkomo"* (Hare-hare we-e we've taken back the land, Hare-hare we-e with Nkomo).[29]

The children's appropriation of Nkomo and Chibwechitedza in their games was part of a larger economy of children's extensions of indigenous knowledge to the changing realities they lived in. They assembled clay and molded it, stripped wire from neighbors' fences to make toy cars like the Citroën, Zephyr, Renault, and "Pijoti" (Peugeot), and anointed themselves Joshua Nkomos. Through *mahumbwe*, complete with its division of labor in the household and the requisite gendered mannerisms, children took command of family affairs (cf. Matibe 2009, 15). These roles were extended to act out Nkomo and his *masaramusi* at Gonakudzingwa.

In the late winter afternoon, the boys and the girls returned from the different recreational sites and prepared to go to *chinyambela* (evening dance to the drum), where one of the popular songs was "Tsuro Tsuro." The boys would sing, drum, and dance all the way to where the girls stood, and the girls would do the same, and then the two groups would become one and dance until midnight. There was no sexual activity involved; there were stiff societal penalties, and respect for taboos was beyond question.[30] It was difficult to capture the lyrics unless one was really close. Children were popularizing the struggle, if only subconsciously, but they were aware that "Tsuro Tsuro" could not be sung during the day.[31]

In the schools, political discourse flourished. Schools like Masivamele Primary School (Sub A to Standard 3) and Muhlanguleni (Standard 4–6) were built out of pole and clay mud. Muhlanguleni was a boarding school near Boli, in Chief Chilonga's area. Pupils from far away built and stayed in grass huts/shacks just by the teachers' huts. The classrooms and houses were built in typical chiTshangana architecture: *simbiri* (ironwood) poles put into the ground, narrow and flexible poles or *tihlanga* (reeds) basket-woven around them, covered with clay mud, with a grass-thatched roof on top. These structures were only big enough for one to stretch the body. The school infrastructure was like this right up to 1980.[32]

Still, the school remained a site for the engineering of African children into disciplined (read subservient) colonial subjects. Children were made to recite racist poems like "Bongwi the Baboon," sing "Christopher Columbus was a Great Man" who went to America in a sauce pan, serenade David Livingstone, the "great man" who "discovered" the *Mosi oa Tunya* (Victoria Falls), and understand the *valungu* "truth" that the black person had no history (Chinodya 1989). Official games like "PT" (physical training), soccer (boys), netball (girls), running or athletics (boys and girls), high, long, and triple jump, and choir or singing comprised the staple of inter-house (intra-school) and inter-school competitions.[33]

As a social engineering tool, "Native or Bantu Education" was designed for the "uncivilized" (blacks), and to mark their differences from the "civilized" (whites). In one race, a child was born uncivilized; in another, a child was born civilized. Education had to be beaten into the uncivilized, hence corporal punishment was applied in black schools instead of persuasive incentive in white schools. The white education officers encouraged physical assaults of black children in the name of discipline. Colonialized as well as patriarchal parents encouraged it (Matibe 2009, 43–4). This was a substantial difference with indigenous knowledge spaces like *huvo, chinyambela,* and *dhema,* which never died, but populated the time between school days.

Hunting in a Turbulent Political Time

Inevitably, hunting would not be the exception to the general politicization. Records maintained at Gonarezhou's southern headquarters, Mabalauta Field Station, indicate that the period from 1965 to 1980 was one of sustained poaching in Gonarezhou. A survey of the content of anti-poaching operational reports collected during and after Chibwechitedza's tenure exposes an intriguing mixture of longstanding precolonial as well as incoming tools of the hunt. Together, these materials suggest a professoriate of the hunt continuing to be organized around traditions, but flexible toward and taming incoming repertoires and adapting to the political situation of its time.

The predominant weapon for small and big animals was the snare, made either from cable or from high tensile wire. The numbers each *mupocha* could set was staggering. One snarer was caught in 1966 in possession of eighteen set snares and thirty-three ready for setting. He had already killed one eland, one buffalo, one nyala, one kudu, one zebra, one warthog, two waterbuck, one elephant, and one crested guinea fowl.[34]

Bows and poisoned arrows were still very much part of the arsenal. For example, in 1968, a hunter named Matsilele Yingwani Hlengani of Pahlela village, who operated in the Sengwe area near No. 4 Cattle Dipping Tank, used bows and arrows together with a flintlock and snares. These arrows were, as usual, laced with *vutsulu* poison that each hunter made. The arrowheads continued to be made not only out of the *simbiri* (ironwood) hardwood, but also from iron ore obtained from the local ironsmiths, who now used scrap metal since Europeans had taken over their mines. Those that had guns still deferred to bows and arrows for a silent killing method, in contrast to guns which, when fired, made a "report."[35]

Dogs continued to be weapons of the hunt and were regularly shot when caught chasing in the park. Like in the olden times they were fed with medicines to sharpen their sense of smell (*hwema*) and to be tenacious in mortal combat with animals.[36] Dogs chased to tire the animal and corner it, as well as sniffed it out of the bush for the hunter to kill. Dogs were used in tandem with bows and arrows as well as guns, but they could not be used near traps or snares. The reason is simple: the dogs would be caught in the wire snare or trip the gin trap. It was common to chase anything from hares to antelopes with dogs, but not buffalo, both for the dog's safety and the size of the prey.

Whenever they were caught snaring, hunting with dogs, or armed with bows and arrows, *mapocha* would also be suspected, found in possession of, or lead patrols to a musket or rifle. The data for the 1960s show that different generations of firearms were in use. The most surprising was the eighteenth- and nineteenth-century muzzleloaders, which continued to be found among the local hunting repertoires, albeit in highly modified form in terms of retrofitting, until the 1970s. The guns were preferred to the Martini-Henry rifle, which locals disdained because of its "mighty kick" if one put in too much explosive. Otherwise most hunters used the .303, whose ammunition continued to be available through clandestine means long after that of the Martini-Henry had dried up. Indeed, there was trade in ammunition between the local hunters on tsetse duty—who banked tails and created a surplus of ammunition—their *mapocha* friends would kill animals and turn the tails over to their kin engaged in tsetse-related game elimination in return for ammunition. The same happened when the department switched to the 12-bore double-barreled shotguns. Other guns in use included the 9.3 Mauser, .22, and .410 for smaller animals and birds.[37]

Mapocha had a very efficient weapon maintenance and production system in place. Ironsmiths (including those that were also *mapocha*) were not only repairing guns, but also making versions of them, part of the continuing tradition seen in chapter 3. In the Mafuku area, in 1964 the anti-poaching unit arrested a man named Jasi for possession of a "home-made gun, slugs and detonators."[38] Men like Jasi were highly adaptive *nyanzvi* in practically all aspects of indigenous hunting *nzira*, who adopted new raw materials to their design. Another was Toyola Pahlela of Sengwe, who is referred to in the anti-poaching reports as "a master trap mechanic." Among instruments recovered from his homestead were a muzzle-loader and cable snares, bits and pieces of two to three gin traps, four hammers and two chisels, and some bellows. In addition to repairing his own equipment, Toyola was the *mhizha* to whom other *mapocha* deferred for new weapons or to repair old guns.[39]

Just as we saw with Bvekenya, these *mapocha* ran their own network to monitor the whereabouts of park patrols and animals. Because they were locals, it was easy for uncles who were scouts to advise their nephews in the village when best to come in. In the absence of any kinship links between scouts and *mapocha*, *kupocha* (poaching) was all about time management, which is why the fence and patrol were ineffective as barriers. As an example, Frank Musisinyani of Mahinga village had been successful as a *mupocha* because he usually operated in the Malipati locality "at month-ends when game scouts [went] in for pay."[40] The poaching gangs were organized along village or kinship lines. It was rare to find a hunter who hunted alone, the idea being to pool resources, especially dog packs. Whereas in some cases the *mapocha* were just ordinary villagers of no rank, in villages like Gezani/Pahlela south of Manjinji, the headman Pahlela Mavindhlu himself was the chief culprit. He hunted with his son.[41]

The fences that the state meant as boundaries between villages and the game reserve created new hidden highways for *mapocha*. For example in the Gonakudzingwa African Purchase Area in 1972, an anti-poaching patrol discovered that some individuals were using Farm No. 17 as a hidden butchery. They would enter the park, make a kill and carry the full carcass back onto the farm to avoid detection.[42] One of the men—Chitayi—was the property owner. The other man was Elias Suzwani, who owned a large caliber muzzleloader kept on Chitayi's farm. He admitted to arranging transport to remove carcasses from Gonarezhou, confessing that "numerous game animals were butchered on his property and that meat was given to him." It was presumed that Suzwani received meat in payment for the rifle and transport.[43]

Kinship-related poaching connections also cut across the border between "Rhodesian" maTshangana and "Portuguese" maTshangana. Take the case of Mambawu, a Portuguese African who lived in Tsvuku village in Malvernia, who in the 1960s used to cross into Gonarezhou with dogs and snares. Then there was Casimitu, arrested on March 21, 1968, for possessing a rifle and hunting deep inside the park. Casimitu confessed his reliance on maTshangana kin in Rhodesia as informants and a ready market for *nyama*. In October 1968, a man named Dumazi who lived in Hayisa village on the Rhodesian side, was arrested for possessing three set snares and admitted to killing two nyala. In court, Dumazi admitted to operating with Thomas, a "Portuguese African" who lived in the adjacent village called Gavumente (Government).[44]

The Professoriate and the Transient Workspace of Self-Liberation

Elias Chauke was a teacher at Muhlanguleni during Nkomo's detention at Gonakudzingwa. Not just *another* teacher—he was the most popular staff member with the increasingly militant youths. He endeared himself because of the sense of outrage and insult he felt watching Chibwechitedza and *valungu* calling his father "boy" and treating him like a child in his presence. What was left of a man when he was made to crawl and grovel before a young man who was the age of his own grandson simply because he was a white man? What was left of being a son when the very notion of a father was so thoroughly dismantled in one's face? How could Chibwechitedza deny him his birthright to hunt in the land of his ancestors, and animals that they had nurtured through their own discipline and spirituality?[45] It was in the fertile soils like Chauke that Nkomo's seed of resistance found conditions not only to germinate but also grow.

Whenever the local boys approached Elias Chauke, they would greet him exuberantly with the ZAPU slogan *"Zii, Tiko lahina."* Chauke would admonish them to be quiet. After classes and during the weekends, he would be found in the villages of Masivamele, Chilonga, and Chikombedzi conferring with other men at the *huvo* about politics.[46] Teachers like Chauke inspired students into active involvement in the militarization of the struggle for independence in the 1970s. They walked the talk. As *makomuredhi* guerrillas (from the Marxist-Leninist address 'comrade'), began infiltrating Rhodesia through the area in April 1976, Chauke slipped into Mozambique and enlisted in the Zimbabwe African National Liberation Army (ZANLA), the military arm of the Zimbabwe African National Union (Patriotic Front), which had split from Nkomo's ZAPU in 1963. Solomon Bvekenya and his brother Peter Piet Barnard, sons of John Piet, the son of Bvekenya, became *vanamujibha* (aides to *makomuredhi).* Solomon would subsequently twice unsuccessfully attempt to cross the border, first into Mozambique to join ZANLA and then, via South Africa, into Botswana to join the Zimbabwe People's Revolutionary Army (ZPRA), the military wing of ZAPU. His friend Hlanganani made it into Mozambique but never returned. Another friend, Manase also crossed into Mozambique, completed training at Tembwe, and went for specialized training in North Korea.[47]

Those who crossed into Mozambique fought, but so did those who remained and kept the institutions they sustained intact. Patriarchal structures and communalism formed the foundations upon which military-style hierarchy and socialist collectivism were made standard operating procedure. Locals did not come to their own struggle for liberation as spectators

Figure 7.2
Solomon Bvekenya.
Source: BBP

waiting for salvation from *makomuredhi* while doing absolutely nothing. Once *makomuredhi* entered Rhodesia and reached the villages, a zone of indistinction emerged between "ordinary people" and *makomuredhi.*

The ordinary person had been drafted into active military service, often without realizing it. *Makomuredhi* established a structure of indirect rule—normally associated with British colonial administration—which was in effect virtual military rule throughout the countryside. As fair game, people died as animals and like animals, were buried by the same undertakers—vultures, hyenas, jackals—and their bones left strewn around, barely recognizable from those of animals except to the discerning observer.

Chimurenga as Guided Mobility

Peter Barnard Bvekenya saw *makomuredhi* arriving before they had even arrived. They came to him in a dream—many people coming from the east carrying guns. In the morning he told his wife. He never thought there might be a war coming.

Two days passed. On the third one, April 18, 1976, they came. Many of them. They arrived while the women and children were in the fields since

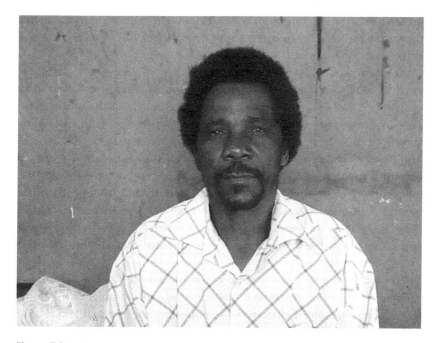

Figure 7.3
Peter Piet Barnard Bvekenya.
Source: BBP

it was during the time of farming. Peter Barnard was at home. They asked his sister Esnat where Peter was. She did not lie to them. He was called there, and he did not refuse. When he saw them, he knew that these were the people the ancestors had introduced to him in a dream.[48] They now introduced themselves—about fifty of them. Peter Barnard remembers five commanders whom he would operate under for a year: Comrades Mutungadura, Makore, Elisha, Kamutsitso, and Metsirapera. They were the first group, highly disciplined, unlike the rogue element that would come after 1978 that broke the cardinal rule of not taking pleasures with the *chimbwido* (female guerilla aides).[49]

The war of liberation was called the *Second Chimurenga*, the first having been the war against colonial occupation in 1896. The original meaning of *chimurenga* is derived from Murenga, the Mwari (God) of vaRozvi; *chiMurenga*, therefore, means the war of Murenga, or war guided by Mwari through his ancestors. As we saw in chapters 1 and 2, all journeying, hunting, fighting, and life itself could only be possible if it was spiritually guided mobility, hence the imperative to pray (*kupira midzimu*) and to ask for guidance (*kutungamirirwa*) when embarking on journeying and to thank the ancestors after arriving safely. ZANLA actually had a *svikiro*-in-residence at its operational bases.

As *makomuredhi* deployed to a new war theater, they received instructions to head to a specific *svikiro* (spirit medium), whose *mhondoro* (lion spirit) would guide their local operations. All throughout the war of independence, *makomuredhi* were believed to possess the supernatural power to "mystically disappear when confronted by the Rhodesian forces" or "change into lions or chameleons . . . or even vanish into thin air." The act of becoming invisible involved calling ancestors to conjure the escape by the procedure of measuring a portion of *buté* [snuff] and placing it on the ground for the ancestors. Then, as the enemy's guns blazed, *komuredhi* beseeched the ancestors to guide him out to safety. Ex-*makomuredhi* will swear that the *mhondoro* transfigured them into forms invisible to the enemy (Daneel 1995).

The role of the spirit (*mudzimu* or *mukwembu*) should not blind us to the role of the *svikiro*, the vehicle, mouthpiece, and container of the spirit. As we saw in chapter 1, the work of connecting the spiritual and mortal realms took place through a real person through whom the spirits arrived (*kusvika*), were consulted, and spoke to the living. During the liberation war the *mhondoro* was the ultimate authority over the fate of not just battles but the war itself. As it turns out, some of the warriors who had seen actual physical combat in the 1896–1897 rising against white oppression now returned as spirits. In some situations, guerrilla commanders became subordinate to their orders (Daneel 1995; Lan 1985).

To fight the colonizer often meant having to live not only like animals in the forest, but also with animals of the forest, sharing caves with dassies, hiding in pools with crocodiles, and sleeping in trees like, sometimes even with, leopards. These are the times when maTshangana and vaShona's notions of animals as spiritual(ized) beings came to the fore, and when life as guided mobility was ever so apparent. Many who survived say the leopards, lions, snakes, crocodiles, and other beasts spared them only because the spirits protected them from harm.[50] *Makomuredhi* who saw combat say they never fought alone, as mere *vanhu venyama* (people-of-flesh), but as spirit(ual) beings, always under the wing of *mhondoro*.

The people of Chibwedziva remember with fondness the guerrilla commander named Weeds Chakarakata, real name Cosmas Gonese, descendent of the Chinamukutu dynasty, who operated in the area for a while.[51] Chakarakata says his ancestor Mabwazhe, a prolific hunter-warrior, would possess his grandson and lead him into battle, to engage in combat with the courage, skill, and tenacity he himself had possessed when fighting past wars (Daneel 1995, 49–105). After killing white soldiers, *makomuredhi* were to abstain from food for a stipulated period, surviving on appetite-suppressing roots and other plant tissue or vitamin tablets when they could find them.[52] To do otherwise was to violate the taboos of war.

We are again confronted by the same unsettling dynamics we met in chapter 1: What happens to skill, technology (guns), or efficacy in the "hands" of the spiritual?

The Many Transient Workplaces Called *Chimurenga*

The *pungwe* was ZANLA's all-night rendezvous with the villagers in the bush or mountains. As an atmosphere, it was populated with singing, dancing, and Marxist-Maoist political education lectures. *Pungwe* was an extension of vaShona's *jenaguru* and maTshangana's *chinyambela* (all-night winter dance) to perform the work of mass mobilization. As a site of comings and goings, it reveals how *makomuredhi* mobilized, organized, and administered the countryside as vast architectures of intelligence, surveillance, and medical, logistic, postal, and transport support to the guerrilla war effort.

Mujibha (youth militia) coordinated intelligence and logistics and were sent on long trips to deliver messages between guerrilla deployments, in addition to transporting and caching ordnance, and engaging in firefights themselves. The name *mujibha* simply described a messenger of the struggle whose task was to be sent *achiongorora zvakawanda* (observing many things).[53] *Mujibha* was also a stage of being groomed for departure to Mozambique to train as a guerrilla. *Kujibhura* (doing *mujibha* work) was a kind of apprenticeship, a transient workspace involving seeing, observing, and reporting, taking orders; carrying messages from one commander to another, risking capture and torture, and certain death; making sure the whole village complied with, was marshaled to, and did not betray the revolution; and, if needs must, taking up the guns of fallen trained comrades in the heat of combat and engaging the enemy.[54]

For Peter Barnard, whom we met elsewhere being schooled in the hunt by his prolific hunter-father John Piet, the war provided a platform for extending his prior knowledge of the forest and the entire area to the work of *mujibha*. As noted, *makomuredhi* arrived in his dreams before their physical arrival, but they chose to come to him in particular, for four reasons, all connected with his father. First, as we saw already, his father John Piet (who had already died in 1972) was not only a great hunter but also a famed *inyanga* (*n'anga* / healer) in Chibwedziva. Second, John Piet had been very loyal to the cause of African nationalism, as attested by his trips to

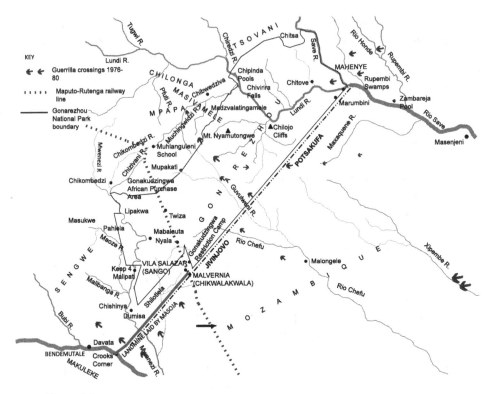

Figure 7.4
Gonarezhou and its environs became venues for transient workspaces of self-liberation from the time Nkomo was detained at Gonakudzingwa to the 1976–1980 war of independence.
Source: Author

Gonakudzingwa to see Nkomo, along with Titus Mukungulushi, Elias Chauke, and other men during the 1960s. Third, John Piet's work as a healer and a hunter had brought him an immense herd of cattle and goats. Finally, Peter Barnard and his brother Solomon were prolific hunters in their own right, having been schooled in the professoriate of the hunt by their father. Therefore, the Bvekenyas were a trusted family spiritually, politically, and logistically.[55]

To say *chimurenga* was guided mobility is to say that *makomuredhi* were not merely spiritually protected in their journeying. It is to also draw attention to how *makomuredhi* deferred to *vanamujibha* (plural) to guide them through the district so that they could then effectively fight the war. In Chibwedziva Peter Barnard was the local guide, executing orders as well as showing *makomuredhi* things of strategic importance in the locality. He spent the rest of 1976 hunting information for *makomuredhi*, showing them around, but basically lying low. In fact, Makore's group was pretty much a reconnaissance or commissariat team to set up structures of logic support and study the terrain. This would pave the way for the combat proper to begin in 1977. Peter Barnard was a convenient choice for a head messenger (*mujibha mukuru*) because he was colored (mixed race, owing to his white grandfather Bvekenya) and the least suspicious, particularly when being sent to spy on white farms and *masoja* (Security Forces) camps. Throughout 1976 *makomuredhi* trained him, sent him on many errands, and took guidanace from him (just as other hunters of the past had required guidance when hunting their quarry in these unfamiliar lands). Only when a close-knit network of *mujibhas* and trusted adult men in the villages had been built did *makomuredhi* then come out of the shadows and call a *pungwe* to introduce themselves.[56]

Among other things, the village was subverted into the quartermaster department of *makomuredhi*, who often carried only their fighting equipment when leaving Mozambique, Botswana, and Zambia for the battlefield. No food. No water. No combat fatigues. Nothing. *Vabereki* (fathers and mothers) ensured that chicken, goat, beef, and in some areas forest meat was always available. *Makomuredhi* did not eat vegetables because, they said, they would develop weak knees and cower in combat.

The energy that drove ZANLA's "the revolution" came from the cooking stick, the sweat, the grinding stones, the hoe, and the slogging feet of *mhaiyo* (mothers) and *chimbwido*, who became the hunters of freedom. All households were expected to take turns to cook for *makomuredhi*. That way, everybody was guilty, and no one could then report others to *masoja*. *Chimbwido* transported the food on her head (another indigenous transportation system and skill) to *makomuredhi*'s bases. The "comrades" ate big plates of *busva*, "as full as those intended for boys going to the circumcision school."[57] As a rule *makomuredhi* did not eat offals, gizzards, hearts, kidneys, or other internal organs, as well as chicken feet, wings, necks, and heads. The reasons were three-fold: (1) they wanted to avoid being poisoned, (2) these meats were considered trivia and undignified for the work of liberation *makomuredhi* were doing, and (3) because each work of industry had taboos and so did the work of crafting an independent nationhood through the barrel of the gun.[58]

In other words, the meat taboos were part of several others representing the extension of age-old taboos to a new kind of hunt: the war pitting Africans against the colonizer, the object of harvest being independence.

Figure 7.5
Komuredhi and *mhaiyo*, two inextricable combatants in the independence war, yet grossly underrepresented in the narrative of Zimbabwe's anti-colonial struggle.
Source: The Zimbabwe News

As in olden times the hunters (*makomuredhi*) were forbidden from engaging in sex (with local women) while on the hunt, if only for reasons of maintaining good rapport with locals. If they disobeyed this and other injunctions as given both by their commanders before deployment into Rhodesia or by *mhondoro* via local *masvikiro*, the ancestors would withdraw from guiding their mobilities and the whole group would fall prey to *masoja*.[59]

As in olden times, it was taboo to disturb restless wandering spirits and bones of fallen enemies. Even today the bones of fallen foe can still be found lying around in places like Gugutsini. It is said if one kicks a bone, it speaks like a person and follows or possesses the kicker who has disturbed this dead man's rest.[60] In Machaze I was severely warned against driving at night along stretches of the Masenjeni to Vilankulo and Chikwarakwara-Shokwe (Chokwe) roads, where bullet- and shrapnel-riddled shells of bombed out vehicles and buildings are still visible. Locals believe that Masenjeni business center, scene of *masoja*'s devastating bombardment in

1978, is haunted to this very day. Not one building has been repaired since. The same story relates to similar sites in Machaze, Chikwarakwara, Mapai, and Mabalane, where "the *fantasma* (Portuguese for "ghosts" or "spirits of the dead") start wandering right after sunset, their lights turning the scene where the person died into day." In Machaze people could not let me proceed with my journey after dark, otherwise "your car will go round and round, stuck in the same place until dawn, and the stomping feet and *'camarada'* (Portuguese for *komuredhi*) singing on top of their voices, going on their battle marches, will be with you all night."[61] I did not dare challenge that advice, and similar advice I had received when touring two other ZANLA military bases, Praça Adriano (Chimoio) and Nyadzonia, in the far north.

Fair Game: Caught between Two Hunters

War is like two hunters hunting each other, and the people are not only the ground upon which the hunting occurs; they are also fair game—people or things considered reasonable targets for military attack, especially given the indistinction between the guerrilla and the civilian population. Attacking civilians is also considered an attack on the enemy; the people become as much part of the combat equation as the guerrillas themselves. Even when their nonmilitary credentials are ascertained, civilians are treated as collateral damage—things considered incidental to the military target.

Watson Machiukele vividly describes the situation confronting the local population in Chibwedziva, caught between these two hunters, *makomuredhi* on one side and *masoja* on the other from 1976 onward:

Anonymous people, dressed as *makomuredhi* but sometimes *masoja aSmith* [Smith's soldiers] dressed in civilian clothes, would ask us: "Who do you prefer to rule you, *vanhu vatema* [blacks] or *varungu* [whites]?" *Makomuredhi* would beat us up badly if we said we want *varungu* since at that time we did not know *kuti munhu mutema unosumudza chii* (what a black man could manage); *masoja* disguised as *makomuredhi* beat us up if we said *"vanhu vatema."* If *valungu* ask whether you have seen *magandanga* [terrorists] and you say yes I have seen them, you will die. If you say to *magandanga* I have seen *mulungu* you will die. So you had to stand and answer in the middle and say: *Anisvitivi*. That answer too might elicit *kupondwa* (severe thrashing with logs). So we learned to say clever things when asked. Did you see *gandanga*? Yes we did. How many were they? How many? They were too many, we could not count them. Which direction were they heading? *Vadai* (They went this way.) Because if you said you have not seen them and yet others have said they saw them, you brought suspicion to yourself. So you had to furnish your story with as much detail as possible, including even offering specifics about what guns *magandanga* were carrying.[62]

In most cases, by their high mobility the guerrillas would be long gone or simply melt away when *masoja's* patrols arrived, rounded up the whole village, and began torturing the ordinary people on the whereabouts of *magandanga* (terrorists). *Masoja* did not wait to attack after the guerrillas had left the village or when *chimbwido* had carried the plates away from the base; they simply opened fire and declared all casualties as terrorists or to have been running with terrorists.

Sometimes *makomuredhi* behaved so recklessly and so unmilitarily that their actions sold the people out to *masoja*. Peter Barnard remembers one particular incident in 1977, when after major operations *makomuredhi* came to his homestead to sit and eat there:

Makomuredhi were in the habit of pulling back into Mozambique after a major gun battle, knowing *masoja* would be fanning the area looking for the culprits and "those running with terrorists." On July 3rd, 1977, the comrades came to base at my homestead in numbers. I said "No, *makomuredhi*, do not stay here because if you do you are selling me out. *Masoja* will come tomorrow and see these tracks. There were many *makomuredhi*, including recruits they were escorting into Mozambique. . . . It was impossible for me to erase the tracks by driving cattle and goats to erase them. This is what we usually did to cover *makomuredhi's* spoor so that *masoja* would not find them.

Then something miraculous happened. That night, July 4th, there was a massive storm. It wiped out all the tracks. That very night I dreamt of many soldiers coming into my homestead, mounted on horses. So I cancelled the journey I had planned to go and see my sister Esnat in Chiredzi, so that the impending torture falls [on] me and not my family because this homestead is mine.

At 7 o'clock they arrived, and I told my mother, Nwa Jeke: "You see, I told you. *Vasvika vanhu vaya*" (Those people have arrived). They rounded up the whole village—children, men, women, young and old, everyone! "Where are they?" They began. "They were here last week," I said. "I saw them. They passed through here heading east, this way. 80 of them." They told me that they had been informed that I was the ringleader of locals running with the terrorists. They pressed me on where the guerrilla camp was, and I said they have no camp, they are ever on the move.[63]

Peter Barnard was taken to one side for interrogation, which always involved beatings. They kicked him in the stomach, on his umbilical cord (and four decades later, on cold days he experiences painful swelling). Then they took him to their camp, handcuffed his hands behind his back, his legs in leg irons, shoved him into a *saga* (sack), tied it up, and threw him into a *handaki* or *gomba* (pit). That was July 4, a Thursday (as Peter Barnard still vividly remembers with stunning accuracy). They starved him for ten days, interspaced with torture, including with electric shock. When he fainted they threw cold July water into his face, continued the interrogations, and resumed the torture until the day's round was up and they threw him back into the pit. On the tenth day, one soldier—a white man—visited the *handaki* at night, gave him a bit of water, and a slice of bread, sternly warning him not to tell anyone. On the eleventh day they took him to jail, and offered him R$300 to work as an informer. He told them: "I do not take blood money" (*mari yeropa*). They tortured him and threw him into the cells for one month and twenty-one days without trial, for choosing to act as a shield to protect *makomuredhi* rather than expose them with information.[64]

Hunting for *Chimurenga*

One day they suddenly took Peter Barnard, placed him in a sack again, completely disoriented, drove for a long time, then dropped him near Chiredzi. While he was at Esnat's house, he was told that *masoja* had also been looking for him in connection with a .375 rifle that he kept to hunt for the terrorists. But he went back to his home anyway, and continued to insist (lie) to *masoja* that it was his Bulawayo-based brother who owned a gun, not him.

Peter Barnard was not alone in hunting for *chimurenga*. In areas around Gonarezhou where animals were abundant, the professoriate of the hunt was called upon to provide meat to energize the knees of the freedom fighters. Stories are told of people who fled into Gonarezhou, from where they set up hunting camps and established meat supply lines for ZANLA. The hunters that had provided meat for their own communities now hunted to provide meat to feed a nationalist guerrilla movement fighting not just for the freedom of the locals, but for all Africans subject to colonial rule.[65]

Rather than being herded into *makipi* (keeps or protected villages) or drafted into the army, some maTshangana fled into Gonarezhou, where they lived until the end of the war. Jephias Mashamba was one of the most prolific hunters in Malipati. Mashamba lived in the Gezani area of Sengwe, and when Malipati Keep was established in 1977 he fled into the mountains inside Gonarezhou's Mabalauta subregion. He survived on hunting to feed himself and *makomuredhi*.[66]

Makipi: Counter-Insurgency as Human Mobility Management

Peter Barnard had been in Chibwedziva for only a few months when *makipi* arrived in November 1977. The success with which *makomuredhi* mobilized people into co-combatants in *chimurenga* had forced the Rhodesian government to resort to forcibly emptying entire villages into *makipi*. *Kipi* was a place surrounded by a wire fence. It was more or less a livestock pen into which *masoja* herded villagers and kept them under duress. They were allocated day times to move out and report back. All persons going out were assigned numbered cards they displayed on their chests, which they surrendered upon return. The number was entered against one's name in the register. Anyone who reported late was interrogated and tortured to reveal what he or she was doing with *magandanga*. Almost five thousand people—all of Masivamele—were relocated to *makipi*.[67]

Chibwedziva business center was one of the many *makipi* that *masoja* established around the villages abutting Gonarezhou. Watson Machiukele was one of those interned in this *kipi*:

They came and told us to go to the *kipi*. We said we are not going to your *kipi*. We don't know what a *kipi* is. At first the *kipi* was supposed to be built at Machiloli but they abandoned the idea because there was a river nearby, and *magandanga* would drink water in that river. So they brought it here at Chibwedziva because there is no water here. They destroyed all the boreholes in the entire area. Having starved us of water, they left us with no option but to come into the *kipi*.[68]

Materially, socially, culturally, and psychologically, the *kipi* represents one of the most dehumanizing aspects of the war. By 1977, an estimated 580,000 Africans had been herded into 203 *makipi* in the northeastern and southeastern "red zones." Where villagers refused to move into *makipi*, *masoja* burned down their homes and crops. No compensation was paid for livestock, homesteads, and crops lost ("Rhodesian Tortures" 1977). Generally, a fifteen-yard square patch was provided for each family "to build a shelter, dig a pit latrine, and accommodate chicken and small livestock. The toilets soon overflowed, unleashing widespread typhoid and diarrhea" ("Health Services at Risk" 1978). Pests were eating people's bodies as they walked. Hospitals were few. People lived like pests—under dusk-to-dawn curfews, with armed guards conducting gate checks on inmates going and returning, "searching their bundles for evidence of assistance to the guerrillas in the bush" ("The Hellish Life" 1978). The only advantage of the *kipi* was the security: in the villages people had been vulnerable to violence both from *masoja* and *magandanga*.

That left the two foes to hunt each other without the villagers. Anyone found in the forest was interpreted to be either a "terrorist" or running errands for "terrorists." The opening statement was gunfire and a dead body. The only terror to Africans came from the cruel *madzakutsaku* (camp guards) ("The Hellish Life" 1978). Still, the guerrillas moved in and out of *makipi*. By 1978 most of them had been abandoned.

Conclusion

As told by locals, the whole "Nkomo at Gonakudzingwa" episode does not conform to the metanarrative of nationalism in Zimbabwe, whereby the political elites engineered the revolution and simply mobilized locals to help them. This is not one of those cases where everybody's spirit of resistance was dormant or nonexistent until the wise nationalists arrived and awoke them. This is what the ZANU elite would like us to believe today, against the grain of those *makomuredhi* who ate the food *mhaiyo* and *mujibha* brought, who survived not through the skin of their teeth but on the wing of the ancestors and the sweat and pain of the common person.

The local struggle was no struggle in the abstract, or simply a national cause; it was the latest episode of a long-running saga pitting a people aggrieved about the loss of their lands and animals against a state that criminalized their access to that heritage. "Poaching" was a critique and a resistance at the individual, family, and community level already under way against restrictions to accessing forest resources critical to survival— like meat, grass (for thatching and pasture), firewood, mushrooms, fruits, and so on. Nkomo only harnessed this resistance, indeed that is just one side of the story, the other being that these locals also harnessed him to their struggle because he was not a black man the whites could ignore without risk. He did not come to their houses in person; they went to him of their own initiative and brought "him" home—as news, in song, as a subject, and a galvanizing force toward their own critiques against and resistance to Chibwechitedza.

When people talked about *hurumende* (government), they did not necessarily mean the cabinet in Salisbury or the prime minister himself, but Chibwechitedza. To make sense, Nkomo had to ground his nationalist rhetoric in the local dynamics of struggle in areas like Chibwedziva, Malipati, and Shilothlela, where Gonarezhou was an issue. In this time when the return to *Mwana wevhu* (child of the soul), African cultural identity was a mobilizing strategy, the local men brought special gifts to the politicians at Gonakudzingwa: skin attire. The colonizer saw Nkomo and his colleagues as criminals disobeying the law of the land; locals saw him as a hero disobeying criminal "laws" of an illegal and racist white minority regime. Who knows? If they worked together with Nkomo, within a much broader struggle against what was apparently a common tormentor—not just one Chibwechitedza but many such cruel oppressors—it might be possible to overcome the seemingly impregnable power that the colonizer possessed.

The professoriate of indigenous knowledge in its various nonhunting and hunting aspects was mobilized toward the project of nonviolent civil disobedience initially and then military resistance later in the 1970s. Politicians seized upon local beliefs like magic as technologies of endearment, recruitment, and incitement to rebellion. It is not clear if they realized that these people were already engaged in a critique of the state as it manifested locally, but they certainly endorsed hunting in Gonarezhou as a birthright—one of many that local maTshangana and vaShona should fight for. Hunting and heritage rights to the forest became a legible example of a more complex assembly of rights that nationalists implored locals to join other Africans nationally in demanding through civil disobedience.

The nationalist message gained traction locally through the social networks of kinship, which served as information highways. From Gonakudzingwa or the rally at Chikombedzi, the word was carried in person to the villages, where it was disseminated at spaces like *padare/huvo*, at beer parties and *nhimbe*, at games like *dhema*, during firewood gathering, at *chinyambela*, at wells, and whispered in the bedrooms between husband and wife at night. The indigenous structures of knowledge production and practice that we saw earlier as professoriate spaces now became vibrant sites of discussion and organizing toward the birth of a new political order. Such were the transient workplaces of anticolonial organizing, places where we seldom look when thinking about nationalism or, for that matter, when describing the work of nationbuilding. Only through attention to the micromobilities of dancing, singing, games, storytelling, and beer drinking are we to find the role of social networks as purveyors of political movements. It was primarily through the family, through relations of kinship, and through the gendered and age-based spaces and modes of socialization that incoming *nzira* (means and ways) of resistance (such as party politics) could find traction.

Remarkably, Africans subverted the structures through which the colonizer had intended to colonialize them and now used them to engage in a process of self-liberation. The restriction camp at Gonakudzingwa became a spiritual site, a shrine where Nkomo performed his *masaramusi* and enlarged the flame of resistance already building within those who undertook the pilgrimage. Schools had been intended to socially engineer a native who would understand just enough of the white man's language and rationalities to properly follow orders, to be a good employee, but never to think critically, to philosophize, to innovate, to be entrepreneurial.

Yet here were teachers like Elias Chauke who defied this conformist agenda even as their fellow members of staff were loyalists and spies reporting their own colleagues, students, and relatives to Chibwechitedza. Far from becoming tools of empire, students under Chauke's custody would follow in his footsteps in leaving the country to go to Mozambique, train, get guns, and cross back into Rhodesia to fight and get rid of the colonizer whom they critiqued as unjust.

Some left to escape a compulsory draft of Africans into the Rhodesian Army, some crossed before there was even talk of any draft, while others were recruited to go and fight. Women cooked for the struggle, men hunted for the struggle; people died for the struggle, while struggling for their own independence and that of their resources. The colonizers had used Africans to fight against tsetse fly; now the nationalists were using locals to fight against the Rhodesians. On each of these occasions locals calculated not according to the national interests of these incoming agents, but their own: meat in the first case (chapter 6), ridding themselves of their immediate (local) tormentors and recovering their lands in the second. The meaning of the national project—be it colonial or postcolonial—depended on the answers it supplied to these immediate questions. The freedom from tsetse fly had remained meaningful only to the white man's economic interests (the cattle ranches) but had ushered in a regime of systematic criminalization of the hunt. Therefore, unless the project of national liberation dealt with the land question in ways that restored local access to Gonarezhou, it would be adjudged incomplete at the very least, or a total betrayal at most.

8 The Professoriate of the Hunt and International Ivory Poaching

He is known as *The Bvekenya of Mahenye*. For much of the colonial period the authorities believed him to be one of Bvekenya's sons with a "Portuguese African" woman. The story of Shadrek Mataruse, the legendary ivory hunter of Mahenye, illustrates the role of kinship networks in his use of guns to harvest ivory from northern Gonarezhou National Park. It would be hard to find anyone in Mahenye who does not know Shadrek, arguably the most famous *mupocha* around Gonarezhou since Bvekenya.

Shadrek lived near the border in Mozambique, from where he made his forays into the park. Barring a period of two years spent on the South African mines, Shadrek had been poaching elephant since 1960. An informer named Puzi told Gonarezhou game scouts that Shadrek had operated with a muzzle-loader in Gonarezhou for fourteen years until the beginning of 1974 when he became more sophisticated. He got arms and ammunition from the same people he sold ivory to: "As he was wounding so many elephant, two European Portuguese, who live[d] at the sawmills on the Save River, about six miles into Mozambique, gave Shadrek a .375 Parker Hale Magnum rifle. These two Europeans are the ones who [bought] his ivory. The name and number of this .375 rifle had been filed off."[1] An informer confessed to the Anti-Poaching Unit [APU] that he had "lost count" of the number of elephant Shadrek had shot, but it was "a tremendous number and that now that he had a .375 rifle, he had collected enough money this year to buy himself a store."[2]

There are two versions of Shadrek Mataruse's story. The first is the one that former Game Warden and Rhodesian Security Forces operative Ron Thomson offers and which says that Shadrek used his ivory poaching activities to spy on ZANLA positions inside Mozambique for Rhodesia. Shadrek, Thomson says, had few options. Prior to Mozambique's independence in 1975 he had an open and thriving market among Portuguese and

Figure 8.1
Shadrek Mataruse.
Source: Safari Press and Sports Afield (with kind permission)

Indian buyers at places like Masenjeni, Machaze, Vilankulo, and Lourenço Marques. The Rhodesians could help, but for a small favor: that as he delivered meat to ZANLA and FRELIMO camps, he could keep a keen eye on their camps, the routines, the mobilities, those comings and goings, and, as he returned to hunt in Gonarezhou, share his stories.

Shadrek was no longer just a poacher of game but of human game as well, a bounty hunter who hunted with his eye and sold information to the fullest extent of *kutengesa* (selling out) possible. The rewards placed on selling out ZANLA commanders operating in the Rio Save area around Masenjeni were fat: R$5,000 for commanders, $1,000 for any "terrorist," hundreds of dollars on a sliding scale for Ruchnoy Pulemyot Degtyaryova (RPD) machine guns, rocket propelled grenade (RPG) rockets and launchers, AK47 assault rifles, land mines, grenades, and ammunition recovered (Thomson 2001, 116–117, 125, 127). The head-turning figures mentioned made ivory hunting and selling look like small change. And this is how the ZANLA and Mozambican army camps at Zambareja (the very one that Bvekenya had hunted from), Madzvalatingamele, Chipungumbira, Garawa, Muzvirizvi, Rupembi (Honde Camp), Chiwi, and Masenjeni—ten camps in total—were attacked in 1979 and destroyed. Shadrek's own hotel was among the buildings destroyed at Masenjeni. He escaped the attack but would be arrested by the Mozambican government and severely tortured. The Rhodesians had apparently abandoned him to his own devices (ibid., 290–292).

Yet not all people remember the Shadrek involved in the national drama. The Shadrek they remember is a man born in Madzvisanga (date unknown), a real professor of the chase who hunted ivory and produced meat to feed his own people. Peers like Elias Joseph grew up with Shadrek; their fathers hunted together and schooled them in the hunt in the same area where Bvekenya had hunted with Matangile and Mazinyani. The division between park and village in Mahenye barely existed; in fact the elephants did not recognize it, and they constantly visited the villages to graze on people's crops. Shadrek would follow them, kill them, and give the meat to the people. They did not see him as a *mupocha* (poacher) in the derogatory way the national parks did; that term began to be used in 1973 when Gonarezhou was extended to Mahenye and people were forcibly removed from their lands.

The war of liberation was already escalating in the northeastern stretch of the border with Mozambique, and people concluded that their forced removal was a security measure to empty out the borderlands and countryside of people who provided *makomuredhi* with food and information. Into this cleft between government and the community stepped Shadrek. The elephants were crossing the Save into Mahenye and destroying people's crops, and up stepped Shadrek to kill these problem animals, take the ivory and leave all the meat for the people. The villagers were quite happy. Here was a man who knew the traditional, historical role of a hunter and that of animals in the society. Every young man was deeply honored to volunteer to go on the hunt with him, to learn from this great hunter who had become a menace to *valungu*.[3]

Shadrek's hunter partner in Mahenye was Elias Joseph. He says that time after time they came close to getting arrested, but Shadrek was a clever man. Sometimes he would wake up suddenly and say the men needed to get moving because intuition—a sense of foreboding—was telling him that national parks rangers were coming to get him. "Often he just knew," Joseph said. Shadrek used to cycle into Gonarezhou from Mozambique, hide his bicycle around Marumbini or in the riverine thickets along the Save, before crossing into Rhodesia, lest the distinctive bicycle tire marks betray him. He knew that game scouts did not patrol that far or cross into Mozambique. He would come into the park, hunt elephant for ivory, and bury it underneath the sandy beds of the flowing Save, then come back to fetch them later. The game scouts would follow his spoor, but they could not find them. Like the hunters of old Shadrek used several baobab trees as hideouts. He would keep his ivory, food, ammunition, and other requirements there. The interiors of these hides were big enough to sleep ten people. He would walk through the bush and not the footpaths to leave no tracks, or tie animal hides on his feet and blur his footprints.[4]

In 1981 he shot two big bull elephants at Chipote, ten miles from Mahenye inside Gonarezhou, and buried the tusks in the road sands. He then approached a white man and made a proposition to sell the tusks. The white man was a game ranger. He agreed to "buy" the ivory. Upon arrival at the cache, Shadrek demanded payment before he could release his treasure. The *mulungu* identified himself and told Shadrek he was under arrest. Unknown to him, other park rangers and scouts had circled the area and were hiding in cover in the bushes. When they saw that Shadrek was overpowering their colleague, they came out, guns cocked, and arrested him. That was to be the beginning of what was supposed to be an eight-year prison sentence.

Mozambican Poachers and the People-Parks Conflict in Postwar Gonarezhou

Shadrek was not hunting alone. The 1980s were possibly the worst period of poaching ever experienced in Gonarezhou's history. The park was caught between two wars. There was the ongoing conflict between the aspirations of the people displaced from it, on the one hand, and those of nature conservationists, on the other. Then there was the war between commercial ivory poachers with international links and shadow networks, on one hand, and the government, the whites who controlled and benefitted from conservation, and international conservation NGOs, on the other.

Just as in the colonial period, Shadrek and other poachers took advantage of the conflict between the Department of National Parks and the community. The wartime politicians and commanders who had made promises were now gone, enjoying their newfound power in the capital Harare. The national park was left to deal with its community. The same critique that African hunters had offered to the colonial arrangement through poaching was extended to this conservation fortress. Shadrek was at the thick of this local critique, but he also represented a phenomenon of convergence between the local professoriate of the hunt and incoming actors.

The wartime promises were not kept. Having been forcibly removed from Gonarezhou during the period of anticolonial resistance (1960–1979),

Figure 8.2
The entrance to Gonarezhou National Park.
Source: BBP

local communities expected a black government to correct this historical wrong as promised. Instead the community members were told that they could not go back; their former lands were now in the national park and they must respect that. Only employees of the national park and the animals could stay. The tragedy of the post-independence government is that it followed the colonial regime in criminalizing the professoriate of the hunt and access to the forest for ordinary people in general. Because of their grievances, it was difficult to mobilize locals to come in as allies to conservation; instead, the partnership that formed was between commercial poachers and local communities in defying the state's ecological choices. The state's financial resources were stretched thin from two directions: from the villages in Zimbabwe and along the border frontage from Mozambique. A more pragmatic option would have been to build a fortress of goodwill among ordinary people so that they could co-own the security of the national park, rather than for nature conservation to be based on a fortress that leaked like a sieve.

The Law as a Refuge for Poachers: Four Instances

The problem with the law- and force-based structure of fortress conservation that the government inherited without critique was that the key pillars that were supposed to sustain it were the weakest parts of the structure. The challenges were at four stages of anti-poaching—arrest criteria, procedures to follow when confronting the poacher, prosecuting him, and giving him a sentence. Overall, the evidence emerging from the archival documents shows clearly that people designated as poachers rationalized the workings of the law, not through reading the document but judging from many encounters with the park's game scouts, the police (prosecuting their cases), and the magistrates (sentencing them). Just as we saw with the tsetse department in chapter 5, the DNPWLM was being pushed to reform itself from below.

Age Limits and Juvenile Poaching

In 1983, Mabalauta Warden R. L. Murray reported that cases of poaching involving juveniles within Mabalauta subregion of the Gonarezhou and Malipati Safari Area were "increasing at an alarming rate." Villagers had figured out that the police took no action against juveniles "and accordingly encourage[d] their children to enter the Parks and Wildlife estate" to hunt. Thus on August 28, 1983, game scouts had apprehended six children from nearby Pahlela who had spent the entire day hunting inside the park with a pack of dogs. By the end of their adventure they had made an impressive harvest: two adult nyala cows, an adult warthog, and a clipspringer.[5]

These youngsters did not calculate the value of the meat according to the Z$ the national park was losing, but according to the values they attached not only to *kuhlota* (hunting) or being *xihloti* (a hunter), but also *nyama* (meat). Hunting was still an educational movement; through it these boys were learning how to become men and future husbands who had responsibilities. One of these responsibilities was to hunt in a forest full of a new kind of lion: the *valungu's* law. For despite the political talk of independence, *valungu* were still in charge and *vanhu* (people, or blacks) occupied very low positions as rangers and scouts. Viewed from the village, therefore, nothing had really changed about Gonarezhou. Thus, over and above the tactics the boys needed to use toward tracking and killing specific animals, adult hunters also taught their children how to be elusive prey against game scouts.[6]

•

The question confronting the Mobile Anti-Poaching Unit (MAPU) was how to deal with underage poachers who could not be given sufficiently deterrent sentences. Who should be charged—the messengers or those who sent them? The Joint Operations Command (JOC) meeting on February 24, 1986, heard from the acting warden for Mabalauta, Senior Ranger P. G. E. Westrop, a tone of disappointed resignation. It was now six years into independence, and the ZRP (Zimbabwe Republic Police) officer at Malipati "had not tried to resolve the matter of juveniles released by Police Mwenezi on a local level." The ZRP Officer Commanding Chiredzi District, Superintendent Chingosho, took umbrage at Westrop's allegations, throwing the matter back at the law and the way national parks staff gathered and packaged the evidence for prosecution. Some of these cases were simply unprosecutable not because his staff was incompetent as Westrop averred, but because the law was flawed.[7]

Clearly, adults of the village had adjusted their hunting traditions to negotiate the vicissitudes of the criminal justice system. Instead of going in themselves to hunt, they were now sending their children in the full knowledge that juveniles were unpunishable under the law. Hunting was spectacularly endemic during the school holidays, "with juvenile poachers . . . causing just as much damage to wildlife as adults." Unlike adults, however, these juveniles were "seemingly above the law and [got] off scot-free." Parents knew that even if their children were arrested, the police would release them without charge and if charged, the courts would let them off "with a meaningless caution."[8]

In August 1985 game scouts had arrested six juveniles of school-going age. Four were sent to Mpakati Police Post. Three of them—James Mbiza, Dhlayani Mapongo, and Chuma Dube—were members of "a juvenile gang of five" caught hunting with dogs in the Malipati Safari Area. Mbiza was the only one above age eighteen, and he was arraigned at Beitbridge court on August 27, 1985. Instead of giving Mbiza a stiffer sentence, the magistrate simply cautioned and discharged him. The other two suspects said they were under the age of sixteen, so police at Mwenezi (formerly Nuanetsi) sternly warned them and "released [them] unconditionally" without going to court.[9]

Interestingly, it was the state and not maTshangana that had instituted jail sentences based on numerical age. Initially the age system was intended for determining the age of taxation. Few people had gone to school; even fewer cared about what date they were born. Age was nothing but a number, often given by *mudzviti* (a magistrate or district commissioner) according to whether one was tall enough to be a given age or "looked" old enough. Age, says Mugocha Julius Mavasa, was measured on a marked vertical wooden bar. This assignment of age by measurement rather than biometrics taken at birth arose from the fact that most children were born at home in the care of *vananyamukuta* (midwives), who issued no birth certificates. The only time most children went into the office of the district commissioner was to get *chitupa* (an identity card). Somebody born at home only went to the hospital to seek treatment (after first consulting *midzimu* and *n'anga* of course), and when they died and were taken to the mortuary and a postmortem had to be performed.[10]

"Juvenile" was a legal category of the state; in the community, the measure of adulthood lay elsewhere. Are you now man enough to be brave enough to make and use your own bow and arrows and *vutsulu*? Are you ready to make and set your own wire snares, train and guide your own dogs tactfully when you are in the forest, locate and track animals until you kill, master the art of knowing where animals and game scouts are at different times of the day and night, and bring home meat and elicit the clan praises from your own mother and even your father—and your wife? Have you a first wife yet? Did you go for circumcision in the forest yet and return feeling like a man—as a man? When the game scouts catch you, do you just panic, wet your pants, and tell them everything, or do you say, "*Anisvitivi*" (I don't know)?

Sometimes "*Anisvitivi*" was not even necessary. The "juveniles" admitted to hunting in the park with dogs to the game scouts then changed their statements in court claiming harassment and duress. Sometimes they admitted and played the state at its own game: the law. The adults that stayed away from the park and sent the youths in their stead did so not only as a tactic to evade jail but also according to the gerontocratic traditions of the professoriate: that of delegating tasks whenever younger ones can execute them to one's satisfaction. There was nothing "juvenile" about the practice because it passed the litmus test of executing the hunt. And yet, under law, the act could not be separated from the actor. Seen from the village, the law—not the specific statute on paper but the educative experience of being arrested by scouts, being taken to the police station and released by police, or being taken to court and released by the magistrate—was conducted on the basis of one thing only: age. This paradoxically equipped maTshangana and vaShona near Gonarezhou with the best way to continue hunting without grave punitive consequences.[11]

Figure 8.3
Even after CAMPFIRE, why does poaching continue? Here, wire snares of different varieties piled up at Mabalauta Field Station. Four handcuffed young men are
paraded with their many bundles of snares, hunting axes, and skulls of their past
kills. The horns show clearly that impala and buffalo were among their kills.
Source: Edson Gandiwa, 2012

Westrop therefore agreed with Ranger Chimanga: if these "juveniles"
could be caned (given a few lashes), and age had consequences instead of
being a protective shield from deterrence, the problem of poaching would
be dramatically reduced. An "example" would be made not only to the
children but the parents as well. Which parents, for the love of meat, would
welcome their son back home with swollen, bloodied buttocks? No—which
child would agree to shoulder personally the pain of looking after his
father's household, let alone that of continuing the knowledge handed
down from generation to generation? As it was, the age shield offered
parents "an incentive to break the laws of the country by sending in underage persons to commit the crime."[12]

Procedures of Confronting a Poacher

How could the game scout be expected to catch the poacher if the law did not allow him to shoot to kill, and yet the poacher did not obey such laws? The law said clearly: "Don't shoot until they fire first." Warden Palmer found this ridiculous: "How do you physically put handcuffs on a man with an AK?" He was seeking "a directive to shoot" instead of the existing procedure, which was "to only fire over poachers' heads." Just as in the case of juveniles and the age loophole, the poachers very quickly realized that none of the lead was being pumped into their fleeing behinds and therefore deployed flight as the weapon of choice: they simply ran across the border into Mozambique.[13]

The DNPWLM Directorate advised Palmer that it had made submissions to the Attorney-General way back in 1976 asking him "to define and clarify the powers of arrest and the manner in which arrests may be made." The result was Opinion No. 23 of 1976 AG's Ref AG/18/11 (m) and Departmental Ref A/15 of 5th May 1976. Another legal opinion was requested on February 24, 1983, to deal with poaching and was obtained on May 3, 1983. It had "little to add" to the previous opinion besides stating that the game scout should apply "acceptable force" to execute an arrest. The two opinions could not explain how rangers and scouts could use firearms to protect themselves and animals from armed poachers. In fact, rangers risked prosecution for wounding or killing a person they saw poaching. What force was considered "acceptable" if it was also "impossible to lay down hard and fast rules on the use of firearms"?[14] Technically, under Statutory Instrument 256 of 1984 and the Parks and Wildlife Act 1975, game rangers and scouts were "Peace Officers" "endowed with capacity to maintain the park's territorial integrity and to preserve the park's natural resources." Rangers and scouts were armed largely for personal protection, and if they injured or killed any trespassers into parks, their liabilities were determined under Section 101 (c) of the Parks and Wildlife Act No. 14 of 1975. The police would "not effect instant arrest and detention of the authority responsible for the injury or death," but would open a docket and investigate the matter fully.[15]

How then would a parks officer avoid arrest himself if he injured or killed a poacher in the process of trying to institute an arrest? In practice, the scout was presumed to be a homicide suspect until he showed evidence that he was not. The bits-and-pieces of statutory instruments and opinions in existence only prevented the officer from getting instantly arrested if he injured or killed a poacher or intruder. It also ensured "there would certainly be an immediate arrest and detention where firearms were used wantonly and indiscriminately." But who really determined "wantonness" if there was no witness to a contact with a poacher whom the scout *had* to kill?[16] Prime Minister Robert Mugabe had warned that government might have "to account for [poachers] in other ways" —in other words, adopt a shoot-to-kill policy. However, Acting DNPWLM Director M. R. Drury sent out an immediate order to all stations, cautioning that the PM's incendiary statement was "not a license to operate in any other way other than that required by law."[17]

At the Level of Prosecution

As far as the Department of National Parks was concerned, besides the law, the weakest link in the anti-poaching architecture was the authority charged with investigating and prosecuting: the police. The wardens were convinced that the Zimbabwe Republic Police was a mass of uncertainty and ignorance when it came to the law. Archives show on the one hand an open conflict between police officers recruited from the ranks of the national liberation movements, and on the other hand the more experienced wardens and rangers over-enthusiastic about conservation, all of whom had served in police and Rhodesian army roles under compulsory call-up during the war.

The police officers, victims of these draconian *valungu* laws under colonialism, were often tempted to think that the Rhodesians had got away with far worse atrocities under colonial rule than killing three guinea fowl. The wardens, all white, had been at the top of the colonial food chain, dispensing violence as BSA Police officers, and they obviously knew how to apply the law to the letter toward the black skin, even while covering up their poaching activities in collusion with South African Defence Force (SADF)'s white command element. Black game scouts and general hands talk of small fixed-wing aircraft and helicopters landing and taking off at Mabalauta airstrip in what they suspect to have been airlifts of ivory.[18] Palmer in particular was prone to "lecturing" the ZRP officers as if they were "subordinates" like olden times. The black policemen in a black majority state did not take too kindly to such "sermons."

The wardens' major accusation was that the ZRP's local officers had secured light sentences instead of more deterrent ones because they were illiterate and did not understand the law. Rangers accused the ZRP of blocking cases from being successfully prosecuted and threatening to cause "the complete eradication of vast areas of Zimbabwe's wildlife . . . in the not-too-distant future."[19] Heavy penalties would "make a statement," but the ZRP was prosecuting cases for ridiculously low sentences. It had long been noticed that besides sending their children in to do the hunting, poachers were also approaching from Mozambique to create the alibi that they were Mozambican rebels and refugees.[20] There was clear evidence that locals were teaming up with their Mozambican relatives to poach ivory in Gonarezhou, crossing back with their bounty. Having left the guns in Mozambique, the Zimbabweans then walked back home like ordinary villagers coming back from a visit to kin.[21]

Even when game scouts had apprehended these culprits, Westrop charged that the police had botched up the process. That is what had happened to another Mozambican poacher named Arimando Samu whom MAPU had captured on December 9, 1984, after shooting and wounding him in the knee in the Nyala area of Mabalauta. The patrol had even recovered an SKS rifle and twenty rounds of ammunition Samu had in his posession, and he should have been charged with illegal hunting in a national park and illegal possession of weapons of war. But the ZRP had kept the detainee at Chiredzi and he was subsequently released without charge. Numerous poaching cases taken to Mpakati "died" there before they even got to prosecution.[22]

Even when further evidence was needed, police were too quick to finalize a case regardless of the result, thereby foreclosing the possibilities of securing a deterrent sentence.[23] When Parks brought a suspect to Mpakati police station, they were assured that "the case is cut and dried and there would be no need for our witnesses to appear in court." The accused pleaded guilty and changed his story in court; instead of asking for the prisoner to be remanded to allow them time to call game scouts as witnesses, the police bulldozed the case through and lost or secured lighter sentences.[24] To make matters worse, the police officers preparing the cases were not up to date with the latest changes in game law and framed charges on the basis of old ones.[25]

The lack of transport was an impediment to the smooth communication between game rangers and the police. DNPWLM did not allocate enough mileage to stations facing serious poaching and that were located far away from the nearest police station. Gonarezhou had been a combat zone during the war. Bridges had been landmined or blown up. Telephone lines were still only found at the main shopping centers like Chikombedzi and at schools like Malipati, but these were not linked to park headquarters inside Gonarezhou. Up until the advent of cell phones in the mid- to late 2000s, real-time communication could only be done by getting into a Land Rover and driving to the police station proper. That same vehicle was needed for patrols in the park.[26]

The ZRP budget was also very thin. The lack of transport was such that even if a case could be won by a simple attendance of game scouts in support of the prosecution's case, non-attendance resulted in insufficient evidence and thus a lighter sentence or a discharge for the suspect. The best hope for successful prosecution lay in bringing the law enforcement mechanism closer to Gonarezhou, or allocating transport to attenuate the tyranny of distance that was the poacher's refuge.[27]

At the Level of Sentencing

If force and jail helped, Risimati Mbanyele would not be perhaps the most jailed hunter in the Chibwedziva area. He is the son of Mukachana Mbanyele, from whom he learned a lot about the hunt. He was born in the area but does not know when he was born. Mbanyele was one of the "juveniles" hunting in Gonarezhou right after independence and, as he became an adult in the 1980s, was in the eyes of the state a "repeat offender." To locals in Bhaule area, that qualifies him as a *muvhimi mukuru* (a most prominent hunter).

Risimati was first arrested for poaching in 1986 and spent one month in jail, which means he was over eighteen years old, the age when one attained adulthood and jailable age in Zimbabwe. His encyclopedic knowledge of the rangers and police officers involved in anti-poaching matches his reading of the forest and ground for signs of animals. He remembers Ranger Zephaniah Muketiwa and police officers Samson Mavasa, Elias Chibuto, and Sergeant Chauke in particular because they used to raid hunters like him at their homes in the dead of night while they slept.

Armed with AK-47 and FN rifles, they would surround the homestead, before one of them approached the door of the bedroom and ordered all occupants to come out with their hands in the air, then arrest the suspect. Risimati was arrested seven times during the MAPU era and given sentences ranging from one to eight months. He also recalls how it was still possible to avoid capture by sleeping at a neighbor's, or out in the fields, or having more than one wife, especially in Mozambique, to which one might escape. When the raiders came, his wife would always say, *"Anisvitivi."*[28]

As early as 1983, DNPWLM had conceded that the fines approach was not working. The director preferred a mechanism of driving a further wedge between the park and the villagers instead of suggesting a formula for harmonizing their interests.[29] The director's complaint was that under Section 6 (1) Parks and Wildlife General Regulations and Section 28 (1) Parks and Wildlife General Regulations 1981 the fines were not an effective deterrent because they were too small. The courts were only empowered to impose fines of Z$10 to Z$40 (Zimbabwe dollar) for killing animals in a national park or two months in jail. He wanted the penalty increased to a level unaffordable by the villagers, pegged at a custodial sentence that was heavy enough for them to feel the pain of killing an animal and discourage repeat offenders.[30]

Afterword: Shadrek, CAMPFIRE, and After

Upon his release in 1982, Shadrek picked up his guns and returned to his hunting grounds. In 1995 he was again arrested and spent the rest of his life in jail until he passed away in 2002.[31] He may have been a scourge to national parks but to his relatives in Chief Mahenye's community, he was a genius, a hero who fed the people—indeed, in my own comparison, a latter day Zimbabwean version of Ned Kelly or Robin Hood, who poached from the park and gave to the people.

The Ned Kelly or Robin Hood figure persisted in Gonarezhou because one program after another launched to engage communities was intended to protect animals, never to address the concerns of villagers. The philosophy of treating animals as endangered species while treating surrounding communities as endangering species became law with the passage of the Wildlife Act in 1972, granting private ownership of animals to individuals. The act stipulated that communities would also have such ownership, but under Rhodesian law, people who lived in communal lands had no title deeds.

Instead, "their" land belonged to the state while whites *owned* the land. In 1978, DNPWLM's whites-only ecology division came up with the Wildlife Industries New Development for All (WINDFALL) program to engage communities living around conservancies on how best to protect wildlife, without addressing the tetchy issue of land restitution. The ideas generated from WINDFALL culminated in CAMPFIRE—Communal Area Management Programme for Indigenous Resources—in 1989. As advanced by its founder Rowan Martin, the program would undertake the development of rural areas through incentive-based conservation of animals. As understood by the ordinary people from what *valungu* said, CAMPFIRE was a program whereby communities would benefit from surrounding game reserves.

Opinion about CAMPFIRE is divided. Some emerging scholarship hails CAMPFIRE as a strong platform on which to build community-controlled and driven sustainable development and conservation programs (Ngwerume and Muchemwa 2011) whose devolutionary moves have long been seen as a critique of centrally controlled fortress ecology (Murphree 1990). The term "responsible tourism" has been attached to CAMPFIRE to highlight its commitment to development issues and not just conservation (Spenceley 2008). The work of Frank Matose (2006) suggests that tying the national park's relevance to "tourism" or to "animals" limits the many other uses of "forests." Indeed, it is seductive to see "game" reserves as the only issue at stake, yet people have forfeited multiple other "beings" that are not counted in any game-based regime of compensation or restitution.

From the perspective of villagers, CAMPFIRE was not a gift from the white-dominated wildlife industry, but an outcome of post-independence confrontations between national park rangers and the local communities. In an effort to seek a solution, DNPWLM approached Malilangwe (formerly Lone Star Ranch) owner Clive Stockill to intercede between Gonarezhou and the locals. Stockill had grown up in the area, spoke fluent chiTshangana, and got on well with local people, who saw in him one of their own, a true white African.

Stockill asked people to open up about what they wanted, their grievances, and their solutions. They were very clear: "We lost our land. We need it back." He said: "No, you can't get it back anymore. It's now a national park. It's only for the animals and only employees can be allowed there, not anybody else." They mentioned the loss of meat. Stockill came up with an idea of allowing the community to utilize any animals that strayed from the park into their villages.

By that he meant that safari hunters would come in and hunt for a fee, take the trophy and leave the carcass (meat) to them. Communities would be shown the money from the elephant or buffalo shot, and they would choose how they wanted the money to be distributed. So communities called each other to a gathering to budget according to their priorities—like clinics, schools, and grinding mills. The surplus money was then distributed in cash to households.[32]

When people started benefitting from CAMPFIRE they withdrew their support from Shadrek and actually sold him out to the authorities. When next he came to Mahenye and asked his usual companions to join him, they tipped off national parks officials. One of the hunters who was with Shadrek and Elias Joseph recalls: "We understood that we might get arrested but we hoped that we would be able to get the job done without being arrested. We certainly did not expect the harsh punishment we got. . . . Shadrek had eight guns and all his things taken and we lost everything. . . . When we got arrested we had everything taken from us. And I got twenty-five years. . . . I had only got five cattle. They were taken and I was jailed for twenty-five years."[33] Shadrek was taken to a jail in a location locals knew nothing of; we now know from Ron Thomson's account that it was Chikurubi Maximum Security Prison. He died in jail. Said his aunt, Nwa Munzumani: "It was difficult to mourn because we didn't know what happened to him or when it happened."[34]

Shadrek's hunting partner Elias Joseph feels the bigger loss was not being able to hunt again. CAMPFIRE was even worse than the establishment of Gonarezhou because now the community betrayed the very same people who could kill problem animals for them. Even when the animal was spotted in the fields, because it brought safari money, it could not be killed. Otherwise the killer would be poaching and would go to jail.[35] Another hunter in Mahenye agrees: "When you grow up being a hunter and that is what you do, it is part of you and you cannot just give it up—so we just carried on."[36]

Like its predecessors, CAMPFIRE's objective was to save the skins of animals for the profit of the predominantly white-owned ecotourism industry while throwing occasional pittances to Africans masquerading as incentives. It evaded the key issue: the unresolved question of forced removals from Gonarezhou. The interest was still that of letting the animals—especially elephants—multiply. However, people were now reduced to the level of spectators, watching the animals and rich, white, and gun-toting elites from overseas shoot them, while they themselves were banned and wired off the game conservancies. They remained poachers.

The tradition of congregating at the carcass after hunters have shot a trophy elephant constitutes the signature of CAMPFIRE today. In the past the ringing shot used to announce the falling of a tusker, sending people running toward the sound. Word of mouth was used to spread news of a kill. Today there is cell phone reception, and word is spread through text messaging either a week or days before the safari hunters arrive, or seconds after an elephant or buffalo has breathed its last. When an elephant is shot, all the meat *is given to the community*. (In other words, "You said you wanted meat, so here it is; hunting itself is unnecessary.")

The tusks and skin belong to the hunter. ("No chief's right to the ground tusk anymore, this is CAMPFIRE.") Instead, only the head, neck, heart, kidneys, liver, and trunk are reserved for the chief's family. The meat is hung out on a wire, out of reach of dogs, to dry. Caephus Chauke, a member of the Mahenye royal family, is more than happy to get four or five elephants per year in meat. Such a harvest "also assures a lot of rain"; by contrast, the ancestors will not be happy if no elephant is killed and meat distributed to the people during the year. In addition, the money from the ivory and skins is given to the Rural District Council, which takes its share of 45 percent. The rest is surrendered to the community.[37]

Thomas Mutombeni, an employee in one of the safari lodges as a tour guide, looks back at what his uncle Shadrek was doing and says if he lived in that time before CAMPFIRE, and faced the kinds of choices and challenges Shadrek faced, he would support what he did. It was only 1993 when Mahenye Safari Lodge was opened (on an island on the Save River, overlooking Gonarezhou National Park) that Mutombeni began to understand that the interests of government and local communities could in fact be mutually productive. The lodge employs about forty people at any given time; four of them are guides like Muthombeni, double the number when the lodge is fully operational. The whole village has been *bought* over to a tourism view of Gonarezhou forest and its animals.[38]

Supporters of CAMPFIRE like Muthombeni ignore one thing: the program and the revenues will keep coming so long as the economies of the Global North—specifically the United States, England, Germany, and Japan where the trophy hunters come from—are thriving. When there is an economic slowdown, discretionary spending falls, and fewer to no hunters come. And, as the economic and political debacle between 2000 and 2008 showed, these heavily donor-subsidized programs work wonders so long as African countries stick to a politics that favors powerful countries where the money comes from. When they do not, the money flies away (Balint and Mashinya 2008). Unless CAMPFIRE promotes local hunting to raise similar and perhaps even more revenue from wildlife utilization, the whole program will become hostage to forces beyond its control. The project is also vulnerable to the decisions of CITES (Convention on International Trade in Endangered Species), such as the one in 1989 when the body imposed a moratorium on all ivory sales. Zimbabwe's stockpile of ivory piled even higher, the elephant population soared, and habitat depletion got so serious that elephants began to travel far into the countryside destroying people's gardens and fields. In any case, for all its positives, CAMPFIRE does not deal with the issue of redressing historical wrongs, and it is designed to push people out of the park so that they can hunt in the communal lands. These lands, under law, do not even belong to them.

Western NGOs that criticize CAMPFIRE say that elephants should not be killed for economic purposes. Other critics say the elephant population will be substantially reduced. Those in favor of CAMPFIRE tell these critics to please mind their own animals and leave them alone. They accuse these Western organizations of caring more for animals than people, that for them conservation trumps the lives and livelihoods of human beings (Sugal 1997, 9).

In 2002, the governments of Zimbabwe, South Africa, and Mozambique signed an agreement to join Gonarezhou National Park, Kruger National Park, and adjacent areas of Mozambique into the Great Limpopo Transfrontier Conservancy. The idea was to create one vast transboundary area stretching from Malelane in the south all the way to Mahenye in the north, and from Malipati and Phalaborwa in the west to Banhine and Zinhave in the east. The animals would be able to roam freely without any human interruptions, and tourists from all over the world would come and enjoy this rich biodiversity, the jewel of the Southern African Development Community. The philosophy underpinning this project—and its funders, none of whom are Africans—testifies to the continuing aesthetic view of nature as a pleasure for the eye and an instrument for stocking the egos of rich white men from the Northern Hemisphere, many of whom staunchly believe in and defend gun ownership and hunting as part of their American heritage. It is upon the altar of their egos that a transient workplace that used to function as a professoriate for educating African children on forest ecology, and a philosophy that treated the forest as a sacred space, have now been sacrificed to the margins.

Meanwhile, despite the laws, the police, the prisons, and the further militarization of game reserves in the region, ordinary people continue to use their knowledge toward hunting outside the law.

Conclusions: Transient Workspaces in Times of Crisis

By 2000, the unemployment rate in Zimbabwe was angling toward 70 percent; today it is put at over 80 percent. So how do people in a country with an unemployment rate of over 80 percent survive? The first and most obvious answer is the diaspora (over two million according to estimates), contributing no less than US$300 million every year in remittances to relatives. Most of these people are in South Africa, followed by the United Kingdom, the United States, Australia, Botswana, Mozambique, Namibia, and other African countries. There is no doubt that the diaspora is a strong economic constituency without which Zimbabwe might not have survived in the wake of the almost complete collapse of the manufacturing sector and the end of balance of payments and other support from the outside.

This situation was a culmination of a series of events that began at independence. Faced with a population that had been systematically deprived of self-empowering education, health, roads, and other essential infrastructure under colonialism, in 1980 the government of Robert Mugabe's party ZANU (PF) had embarked on a robust Socialism-anchored policy of government-subsidized free education, free health, social welfare, and road construction. Some of these programs were paid for with donor assistance; many of the programs were funded from the national budget. This social support structure, while unavoidable for a liberation movement government, was economically unsustainable in the long term. In 1990, Mugabe reluctantly gave up his single most important achievement—a free health, education, and social welfare program—and agreed to the International Monetary Fund (IMF) and World Bank's economic structural adjustment program (ESAP) that removed most of the social responsibilities of the government to the people.

The words "formal sector" and "informal sector" were always in the news as we grew up in Zimbabwe in the 1990s. They reflected what was becoming a gradual erosion of the "formal sector"—the manufacturing industry formed during the colonial era and still owned by whites. As the IMF and World Bank–prescribed austerity policies cut subsidies to every sector, the government liberalized the economy to make it open to foreign investors.

Like poaching, the rise of the informal sector also served as a critique (by ordinary people, through their actions, their mobilities) of top-down policies: top-down at the global level, top-down at the national level, and top-down book economics that did not articulate with the ordinary lives of people. The informal sector spoke against regimes of top-down planning: its very existence was a statement against linear thinking. Here was a situation in which the Bretton Woods institutions arrived with a linear model of how economies can be "rationalized," technology transfer effected from the North into Zimbabwe, and companies trimmed to a bare minimum of employees.

But the problem did not start with the IMF. Even before independence, those who would become Zimbabwe's leaders had adopted a Marxist-Socialist agenda that criminalized—politically speaking—all indigenous modes of leadership and governance as feudalist. The agenda for revisiting indigenous knowledge was already compromised by the nationalist government's modernist, top-down approach, which staunchly believed in technology transfer. For all its oppression, the one thing one can only admire about the Rhodesian regime was its "import substitution" program, under which it survived over a decade under sanctions through indigenous innovations, particularly in the manufacturing sector. The whole range of armored cars retrofitted onto civilian vehicle chassis is a classic example. But most such projects in military and civilian manufacturing were not continued after independence. Instead, the government pushed the sort of agenda captured well in the United Nations Conference on Trade and Development report in 1972: "The acceleration of the rate of economic growth of the developing countries and the rapid improvement in their social structures through the eradication of mass poverty and of inequality of income require *inter alia* a large-scale transfer from the vast pool of technology accumulated mainly in the developed countries" (UNCTAD 1972, 1).

By this token, Zimbabweans became merely users consuming Northern technologies; they did not even have the power to "modify, domesticate, design, reconfigure, and resist technologies" any more (Oudshoorn and Pinch 2003, 1). The top-down development approach divested policy attention from the creativities of citizens and focused it on the movement of development and technology to them, perpetuating an addiction that could only be satisfied by the outside world. Consequently, Zimbabwe the user became a perennial dependent even for things the country could do for itself. With ESAP, the onslaught on indigenous solutions was intensified.

Although Mugabe abandoned ESAP in 1995, the damage had already been done. The floating of the Zimbabwe dollar, once at parity with the British pound and twice as strong as the United States dollar at independence, weakened the currency and drove prices up. Cheaper imports choked local manufacturing; with demand reduced, the industries retrenched thousands of workers to break even. Food riots erupted in 1997.

The economic situation worsened in 2000 when the government finally decided to deal with the land question. Since 1980, 80 percent of the land was in the hands of less than 1 percent of the population (whites), and few blacks questioned the necessity for land reform. (In fact, the grassroots complained that the government had taken too long to resolve this emotional issue, but the politicians ignored them or said their hands were tied by the Lancaster House peace agreement that ended the guerilla war). Now ZANU (PF) also used it as a weapon against its opponents: voting the party that had fought for the return of the land out of power was tantamount to returning the country to its former colonizer, Britain, which sympathized with the opposition. In the ensuing economic hardship and violence, an estimated three million citizens fled the country into the diaspora. Some directly involved in political activity fled violence and potential for victimization. Many more realized that if they remained in Zimbabwe, their families might starve to death. Just as their forefathers had done, these women and men migrated out of reach of famine as a strategy of *kunosunza*—going to lands afar to hunt or to work in return for food and other livelihood tools, and sending them home ever since.

However, not every Zimbabwean has relatives abroad, so how have they eked out a living in an economy with an unemployment rate of over 80 percent? To address this question, we must first reject the dualism whereby formal sector careers count as employment while the vast array of mobilities that animate ordinary people's entrepreneurship in the informal sector do not. Once this bifurcation is removed, the better question becomes one of examining what kinds of work people are engaged in.

Here we return to the need to study the mobilities that have animated the quest for survival in the most difficult times in Zimbabwe. One of the most remarkable things that happened in the period during and after ESAP was how people returned to modes of innovation and entrepreneurship criminalized under colonialism to eke out survival in an increasingly harsh economic environment. The most obvious feature of this informalization—a decolonializing move involving ordinary people—was the breaking down of the linear boundaries that the government had defined in concrete, fence, language, and law. That is what formal meant: things that fell within the grids defined by the state, or "the modern" (i.e., mostly things of outside origin or derived from the West). By contrast the informal was defined as the "primitive," the opposite of the modern. These are not categories that ordinary people who innovate and deploy strategies of livelihood and creativity use; I will not use them because that is not what careful observation and experience has shown me. The informal sector is therefore simply indigenous creativities upon which the majority depends (Fyle 1987, 498–499).

The concept of "unemployment" often implies that people are loafing or idle. The discourse coming out of Zimbabwe now, but long a fact of everyday life, is that a lot more employment of mind and body to productive activity is happening outside the formal sector in ways that Western-derived bookish economics cannot account for. This everyday innovation is most apparent when a country's economy is in a tailspin, when everyone except those whose bodies are physically impaired are engaged in constant movement. This book has tried to illustrate this from the colonial borderlands, but it also applies beyond. In the ESAP period and since, vibrant "informal" structures surfaced throughout the country. In towns, every place became a potential marketplace.

Livelihood strategies marked by transience (here now, gone the next moment) became the order of the day—at street corners, school gates, street lights, at bus terminals, along the roads, in the undersides of bridges that are empty during the day and bedrooms for "street kids" by night; overcrowded places like Mbare Musika (the country's biggest marketplace and bus station) teeming with vendors, travelers, and pickpockets; carpenter's workshop with workbench, clamps, saws, planers, and shiny furniture by day, just another street corner or pavement by night; and so on. Someone just takes a dish full of "sweets" (candy), cigarettes, matches, tomatoes, muriwo (collard greens), boiled maize cobs, boiled eggs and salt folded in small brown paper "like weed" (marijuana), mazitye (crumbled secondhand clothing from China), sits by the road or streetside, turning it into a transient marketplace. Women and men travel to Maputo, Beira, Johannesburg, and Durban to buy goods, spending as much time on the bus as they spend at home, so that bus and house become transient workspaces and homes folded into one.

Touts shouting themselves hoarse outside to woo customers into the shop; retrenched urbanites retiring to the rural areas *kunorima magadheni* (to grow vegetables in wetland gardens) and take them to the city in *misafa* (bails of recycled synthetic bags to sell); others, back in the villages, resorting to *hupocha, humbavha* (robberies), and cattle rustling. Bustling roadside markets, outward hints of industries that supply them; sculpture and the sculptors of stone and wood; beautifully decorated "traditional" clay pots and the women potters; *mazhanje* (wild loquat), wild mushrooms like *chihombiro, nhedzi, huvhe, uzutwe* and *tsvuketsvuke*, and the mothers, girls and boys that travel miles to pick them, often risking arrest and physical assault from the white farm owner's African guards; sweet potatoes, vegetables, and beans and the women and children (and increasingly men fired from their town jobs) who grow them.

In these creativities, the family unit and the village are the core structure of innovation. All the villagers pool their cattle into one herd, then take turns to herd them, each household for two days. That way, each family in a village of thirty households herds the cattle for only two days every two months, as opposed to every day, thereby freeing time for other work. *Madzoro emombe* (pooled cattle herding) is one of the transient workspaces in rural Zimbabwe today that exemplifies labor maximization through the mobilization of kinship and its *nhimbe*-like capacity to become mass labor. It takes place in *mapani* (valleys) or *mafuro* (pastures). We have seen the second site, *sango* (forest), and the transient workspace of the hunt and its many temporalities. A third site is the *rwizi* (river), with the pools as transient workspaces for the production of *hové* (fish), for consumption or for sale. The fourth is the *gadheni* (garden), where each family grows vegetables, involving constant trips in and out of the shallow well, water can in hand, watering beds of tomatoes, *muriwo* or *muroho* (leaf vegetables), *hanyanisi* (onion), and other relish. And finally, the highland fields (*minda*), where the family grows the staple food for the *dura* (silo or granary) that sustains it for the whole year.

It is impossible to locate these sites of innovation or to define them as technological outside the social tapestries within which they are produced and the bonds to which they bear witness. Whether it is gardening, party political activism, churches, *mabhawa* (bars), schools, fields, *madhibhi* (dipping tanks), clinics, *magrosa* or *mastoro* (grocery stores), villages themselves, local NGOs, *zvigayo* (grinding mills), *vakanyi vezvidhinha* (brickmakers), *mhizha* (steelworkers), or *vanoveza* (carpenters)—all these nodes of technological activity cannot function without kinship and social relations. The local party chairperson or ward councilor, pastor or church elder, gardener, bartender, headmaster, *mudhibhisi* (dipping tank attendant), nurse, storekeeper, villager, aid worker, *magayisa* (miller), brickmaker, carpenter—these people are what defines the intellectual pool available in rural Africa.

Each pursuit exemplifies mobility as work and the optimization of labor value through multi-tasking. A mother going to the garden carrying a hoe or can, spends the day staving off cattle from entering and eating the vegetables in her garden, engaged in weeding the crops, watering the vegetables, and making beds for planting. The party activist who walks or cycles from village to village, rallying relatives to the party meeting, creates a platform for the politician coming from the city to spread certain visions of governance, including technocratic ones. The church pastor trying to build a congregation targets the mother of the house, knowing full well that the responsibility for the spiritual direction of the family rests upon women.

Sure enough, every Saturday the Seventh Day Adventist Church's *Madzimai eDhoka* (Women of Dorcas), in their green and white uniforms of faith, pull their young sons and daughters along *nzira* to the church, bibles, pamphlets, and hymnbooks in hand. Come Sunday and it is the white hats, white collars, red jackets, and black skirts of *Vakadzi Veruwadzano* (women of the congregation) of the Wesleyan Methodist Church. It is very clear: it's Sunday today.

Otherwise those children who do not attend church spend Saturday working in the fields with their parents—weeding, plowing, and harvesting. Typically, most men are at the local *bhawa* (bar), drinking what used to be Chibuku, but is now called Scud, since its container's shape resembles a Scud missile. Chibuku is the opaque beer that we saw in chapters 1 and 2 being brewed by women, which the colonizer found profitable to mass produce in urban factories and sell to Africans, while criminalizing the home-based brewing itself as "illicit" liquor. Just like hunting, the practice continued to flourish, and so did the despiritualized *bhawa*, where men—and local prostitutes—gathered in the evenings or during weekends.

Typically, the children take over chores from their parents as soon as they arrive from school or during weekends. Those whose family have *dzoro* will be herding the village herd, trawling the valleys for spots filled with green grass, mindful of the animals straying to eat crops in the fields, mixing up with the herds of other villages, and getting lost. In the late afternoon, around 3:00 p.m., when the herds have grazed and drunk to the fullness of their stomachs, the youthful herders drive the cattle to a spot where boys rendezvous to play the most popular sport in Africa: soccer.

In times past, the children used to play *nkhati* or *dhema* when at school, coming from school, and after school, but now *vangasvitivi* (they don't know it anymore). British colonialism, specifically the introduction of soccer, killed the game. Now only elders know the game; perhaps, with the growing interest in cricket (another British sport so similar to *dhema*) the sport will come back. The art of children making their own play-balls is still very much alive. Given that soccer balls are expensive to buy, most children in rural Africa make their own soccer balls by stuffing as much plastic, foam rubber, and other "bouncy" material into thick plastic or cloth, tie the end, flip the plastic to create another layer of plastic skin, then tie the end with rubber band. Growing up in the 1980s and 1990s, I remember we used to go a step further: to get rid of the annoying rubber-tied protuberance, we would weave a net around the ball. It would be very durable, and it could go a long way in the sandy and thorny countryside, where normal soccer balls succumb to wear and tear and thorns after only a few matches.[1]

Rhythms of movement shift in midweek. The boys are woken up early in the morning, around 3:00 a.m., to plow or weed the fields. Come sunrise and their parents let them off to go and bathe and commence the long run to school on foot. It used to be barefoot until one reached secondary school, eight years since Grade 1. The school is five to ten miles away for most, a stone's throw away for some. Between 6:30 and 8:30 a.m. the countryside is littered with blue, red, black, brown, and other colors of small "ants" following their tiny *nzira*, separate at first, leaving homesteads, then the village *nzira*, then the main one from the village to the school. Meanwhile the parents are in the fields, hoeing away, watering cattle at the pool, herding in the valleys. Then from 4:00 p.m. the reverse mobilities of schoolchildren returning home happen until the cows come home.

Wednesday it is *zuva rekudhibhisa mombe* (cattle-dipping day), and it is the responsibility of the boys to take the stock to *dhibhi* (dipping tank) to plunge into the pesticides and get rid of ticks and other parasites. Those coming from far away wake up at 3:00 a.m. and are first at the entrance of the cattle holding area. *Sadhibhi* (the one in charge of *dhibhi*) arrives at 5:00 a.m.; the *samvura* (pump attendant) having the day before pumped the water into the plunge-pool and fed the necessary "dip" chemicals, the *mudhibhisi* gets his cattle registers out, sanctions the first comer to get the cattle in. The heifer reluctantly passes through the narrow entrance to the jump-off point, hesitates and in fact tries to retreat, but the herder prods it in the buttocks. It takes the plunge, is completely submerged for a split second, before it emerges, head first, and swims toward the shallow end that rises at 30 degrees from the deep end. It emerges from the arsenic-filled pool, and walks through a narrow corral into the *danga* (holding area), where all the cattle are held until the chemical-filled water drips out of their body.

As the cattle dip, the stockowner's cards are marked against the cattle register. How many bulls, how many cows, oxen, heifers, and calves have dipped? How many cattle are absent, have been sold or bought, have died, or been born since the last dipping day? Any illnesses since? The process complete, the dripping chemical-filled water having stopped and no longer an environmental pollutant, *sadhibhi* authorizes the cattle to be exited. The boy drives his herd home at a trot. Moments later, he is clad in a blue shirt, gray shorts, and running to school barefoot.

It is 2008. Zimbabwe is caught in its worst economic crisis. There are no medicines. There is no fuel. Most of the country's doctors and nurses have joined millions of other Zimbabweans fleeing the economic crisis and political violence. The AIDS pandemic is claiming the lives of many, who die silently for lack of medicines. Somewhere along the path in a country-side called Chihota, an elderly couple is leading out a span of oxen towing a scotch-cart. They are taking their daughter, struck by the HIV/AIDS virus, to the nearby clinic. There are no motor vehicle ambulances here; the cart is THE ambulance. They arrive at the clinic. The one remaining nurse—most of her colleagues having migrated to the United Kingdom—is cheerful, but that is all she can offer these parents. Her face expresses the painful emotions of failing to deliver the help she joined the service to offer—Hippocratic Oath and all. Even the Paracetamol painkiller she used to give such patients has run out. In a choking voice she says: "*Ndine hurombo baba naamai. Motoenda naye kuJenarari kuMarondera. Handina mushonga*" (Sorry father and mother. You will have to take her to Marondera General Hospital. I have no medicine). The foreign currency situation has crippled fuel imports; the petrol shortage has crippled rural transport. The only transport available is the two oxen and the scotch-cart. The general hospital is forty miles away. The choice is clear.

The scotch-cart is not the only default technology that Zimbabweans reverted to when the state of the art failed. Despite the fact that European settlers' first urban and long-distance transport was the ox-drawn cart, by the 1960s a cart arriving at Marandellas General Hospital (as it was called then) would have been deemed a pollutant and drawn instant arrest. How would it get into a hospital located within this "whites only" area? Who would allow it to trudge along De Jager Avenue, let alone through downtown, or from the east through Paradise Park?

And yet the same technologies that the colonizer had criminalized now came to fill the void that those that *valungu* had legitimized and imposed were leaving. This was not necessarily a case of "the shock of the old," where "old" technologies were complementing the "new" or even replacing them (if only temporarily), as Edgerton (2007) suggested. Carts have been the means of transport in the Zimbabwean countryside at least since the 1880s—to cart livestock manure for spreading in fields, for carrying and stockpiling *mashanga* (maize stalks for feeding cattle in winter), for carrying *chibage* (maize) to *chigayo* (grinding mill) and returning with *hupfu* (mealie-meal), for carrying ill relatives to the clinic, and so on.

When people failed to find any treatment for their relatives at local clinics like Border Church, they also failed to find any at Marondera General. They had only two other options for therapy: to seek the help of *n'anga* and traditional herbal medicines, or that of *maprofita* (Christian prophets). Interestingly, *n'anga* and their medicines belong to the domains of endogenous knowledge and practices criminalized after the establishment of colonial rule. The anti-witchcraft legislation passed after the 1896–1897 risings to clamp down on the *masvikiro* who had played such a decisive role in the rebellion, made it criminal under the law to "practice witchcraft" and to be a "witch doctor." However, what the law said and what Africans did were two different things.

Generally, mobilities—including life itself—continued to be guided by the ancestral spirits even among those Africans who claimed to walk the way of Christ under the guidance of the Holy Spirit and toward the Cross. The *bira* (conference with ancestral spirits) was still held; our fathers, members of the MCU (Methodist Church Union), educated at Waddilove Institution, continued to not only attend but also lead preparations and executions of *bira*, travel to Mutiusinazita (Tree without name) to ask for rain from the *mhondoro* Chaminuka, prophet of vaZezuru, to seek answers from and invite *n'anga* to come and cleanse the village and conduct witch hunts, and kneel before *rukuva* (pot stand) every time we were about to embark on a journey, to go to school in Marondera, and to thank the ancestors for our success.

When family members fell ill, the father and mother would take turns to pray to their ancestors, and then after this spiritual stage was done, and the paths to other forms of physical therapy having been purged of *zvimhingamupingi* (obstacles), the patient would be taken to hospital.

When the condition failed to heal, and even while the patient was in hospital, those at home would be busy visiting *n'anga* and *maprofita* to seek reinforcement or alternatives to *vangerengere*'s medicine. Thus when the health system in Zimbabwe collapsed in 2008, newspapers were awash with reports, videos, photographs, and cartoons of people deferring to *n'anga*. What was worse—watching their loved ones die or for them to die *while at least trying something*? Yet it was more than resorting to old ways: Africans have never had just one therapy. Rather, research has shown that the therapeutic itineraries of the ill person have many stops and destinations, none exclusive to the other, all decided upon by the relatives of the ill, who are the therapy managers (Feierman and Janzen 1992). Realizing it was a waste of time to even go to the hospital, people flocked to *n'anga* instead.

Back to 2008, the collapse of public health care and the public transport system went hand in hand with the collapse of the other public services. Energy was one such utility; besides the fuel supply system (Zimbabwe depends entirely on fuel imports), the electricity grid was conspicuous more by its blackouts than power supply. Zimbabwe has not developed any public power supply source since the Kariba Hydroelectric Power Scheme was begun in 1953 and completed seventeen years before independence (Mavhunga 2001). Like most colonial projects, this power was meant for white settler industries and residential areas. The power lines transmitting from Kariba passed overhead to light distant cities, farms, and mines, while the African townships and countryside continued to rely on firewood for cooking, paraffin for heating (Chikowero 2007). Having come to power in 1980, the government embarked on a long-term program of electrification, first in the African townships like *Cherima* (The Place of Darkness) in Marondera that now became *Rujeko* (Light), before moving to the rural areas.

Starting with *magrosa*—since that was where clinics, *zvigayo*, *mastoro*, and other installations that required electricity for refrigeration, milling, lighting, and so on, were based—the program then moved to electrify schools and villages. This rural electrification was made without concomitant expansion in the electricity supply system; the 1,000 megawatts produced at Kariba, plus the 300 at the coal-fired plant at Hwange, Munyati, and Bulawayo, were stressed. Power imports from neighboring Mozambique's Cahora Bassa plant required foreign currency that the country did not have.

Frequent blackouts became the order of the day; power only came for two hours, in the dead of night, and then it was gone for the next twenty-two hours. So people resorted to the means of energy that they used anyway in their rural homes: wood fuel. The city air smelled and looked smoky. The rich and middle classes in the suburbs bought generators; the majority resorted to firewood and charcoal, sourced from Africans who had just received land taken from whites under the land redistribution program that began in 2000. Here, again, people sought solutions in the tried and tested energy forms that have sustained rural life, albeit at immense cost to the forests, since they invented fire.

Water supply too collapsed. Traditionally, African water supply does not come from open rivers or lakes where all dirt—be it vegetative, excrement, or bad spirits—are cast away by means of rainfall and runoff, and through cleansing ceremonies we saw in chapter 1. The well, better yet the spring, was the primary water supply source of the village. Its upkeep was guaranteed through a panoply of rituals and taboos. No black pots were to be used to fill the water pitchers at the well, or else *minyatso yezvisipiti* (the nipples of the springs) would dry up. To be called *nhundiramutsime* (the one who urinates into the well) was the worst of all rebukes. And the mouth of the well was to be covered with the branch of a thorny tree so that the guardians of the pool would prick witches into retreat; in reality, this measure was designed to prevent cattle and goats from drinking. In the countryside this public water supply system continued throughout and after the colonial period despite the increasing construction of *migodhi* (wells thirty feet deep depending on the water table) at individual homesteads. *Kuchera migodhi* (digging wells) was a skill that only one or two men in a village or several villages possessed.

In well-watered areas, a shallow well two to three meters deep was good enough. But in urban areas, big dams were constructed near each town, "waterworks" were installed to purify the water, trenches dug and pipes laid and buried in the ground to carry the water, initially to supply the homes of whites, their industries, and public places, and then extended to African townships. But in 2008, the vulnerability of this centralized water supply system was exposed when foreign currency to import water treatment chemicals became scarce. Raw untreated water was simply pumped into people's homes. The affluent ones drilled boreholes in their backyard—indeed an urban incarnation of the deep well so typical of the rural countryside. Again, as with firewood, the rural was coming to the city. Even the raw water became very scarce because there was no electric power to pump it; so while the rich deferred to boreholes, ordinary people in the townships teamed up as neighbors to dig shallow and deep wells along the Mukuvisi, Hunyani, and other rivers. Every sunrise and sunset, women and children and bachelors could be seen trooping to the wells, bringing back home water in jerry cans, carted on wheelbarrows, balanced on their heads.

Also in this crisis period, traditional ways of waste disposal came to the fore. In pre-European Zimbabwe, the toilet was generally the forest. *Kumbofambafamba* (just taking a walk) is the word used to describe a trip to pass stool in the forest, far from the public view, hidden in the bushes. Men went in one direction, women in another—the way people still do nowadays when a long-distance bus breaks down in the middle of nowhere. Otherwise going to the toilet was something done in the course of movement, when nature called—during the hunt, on a foot journey, and so on. There were strict injunctions against urinating or defecating in or near sources of drinking water and in areas where edible fruits and mushrooms, shrines, cemeteries, and other designated places where human waste would cause *kusemesa* (messiness).

It is not clear when Africans began using pit latrines, marked distinctly *"vakadzi"* (women) to one side and *"varume"* (men) on the other. What is beyond doubt is that this type of toilet had become the most advanced mode of human waste disposal in the rural areas by the time of independence in 1980. Needless to say it was also for those who could afford it; many more people just went around the corner. The forest system was effective when people went far away from the homestead, and the cover was good. With the forced removal to native reserves, such a system lost two assets that made it hygienic and private: distance and cover. The problem of *kusemesa* and lack of privacy was even more serious in the *chirambahuyo* (slums, literally "the ones who refuse to grind their own grain," or lazy ones), where children and adults just "went around the house" and "offloaded their shit." The situation was even more acute when it rained.

Fast-forward to 2008. The flush toilet system that the colonizer had introduced and socialized the urbanites in relied on a regular supply of water to fill up the cistern and flush after passing stool. The water rationing could go for days before black, gunk-filled squirts announced the agonized trickle. People initially filled buckets, pouring measured quantities after "finishing." But there was another problem. The sewer pipes had been laid in the 1950s and 1960s; it was said that the maps had been lost, and so a pipe had to burst first before it could be repaired. There was no money to buy the piping; the industries were closing because there was no electricity to run the machines to make the pipes to repair the sewage system.

So the human excrement one poured water on and got rid of was simply getting out of the house and onto the surface of the road right outside!

The suburbans were less affected because their houses had been built after independence; they were neighbors to "important people" and could count on repairs being done; or they could pool resources in the same street, send someone over to South Africa to buy the piping and get a handyman to repair the leakage. For those living in townships, especially on the edges of the city, there was often one choice: the forest system.

The practice of farming as a strategy of securing the household's staple food and vegetables also reached a high point in 2008. Throughout the colonial period, it was normal for men to cycle home on bicycles after knocking off from work around 4:00 p.m. on Fridays. Come Sunday late afternoon they hit the road, a ten kilogram bag of maize meal strapped to the carrier, together with green maize cobs, boiled or roasted peanuts, and some green vegetables from the gardens, all of them the fruits of their wives' and children's labors. Even in towns, Africans grew vegetables, maize, and sweet potatoes in their backyards. Yet the scale was too tiny to be called "agriculture"; the urban backyard was, put simply, a garden—and too tiny to compare with the sprawling rural *gadheni*, which was almost as big as the upland fields or the homestead fields four to twenty acres in size.

All that changed dramatically in the years after 2000. Prior to that date milling companies like Blue Ribbon Foods and National Foods had kept the supermarkets awash with factory-milled maize meal. Only those with homes in the countryside had relied on their rural silos for the staple. However, as the government seized white-owned farms a significant source of commercial grain was decapitated. The disruptions to farm production were witnessed in the cities as empty shelves. Maize fields mushroomed on every vacant spot in the city and its peripheries. This peri-urban agriculture complemented and amplified the more sustained pattern where Africans had since colonial times relied upon their rural fields and indigenous farming methods and crops to supplement their urban incomes and diet.

As we saw with African men going to the mines, to enroll as *magocha*, and to work on farms, the city was *vangerengere*'s world, the new forest that required guidance from the ancestors for one to be prosperous. This diaspora—as seen from the clan's point of view in the African countryside—was not a home but a journey from which the individual who was away from home (the village) would return. Some were swallowed, lived with a "city woman," and never returned (*kuchona*, or going for good), returned in old age with nothing, or constantly went back every weekend to their families in the countryside. This diaspora was reachable on foot, on bicycle, and by bus.

On the farms and in the mines and factories Africans coming from the Rhodesian countryside became part of a much larger labor pool. Here, history offers interesting lessons. By 1954, half a million Africans were employed in Southern Rhodesia's mining and agriculture industries. Of these, half were foreigners while half were from Southern Rhodesia itself. Of the foreigners one-quarter, one-sixth, and one-tenth of able-bodied immigrant workers were from Nyasaland (Malawi), Northern Rhodesia (Zambia), and Portuguese East Africa (Mozambique) respectively (Scott 1954, 29). Africans stayed at any given mine only for as long as the needs that had taken them there remained. At the old mines like Wankie and Renco Africans were becoming "temporarily urbanized" or "detribalized natives." On tobacco and maize farms, cattle ranches, and sawmills, most were short-term workers (ibid., 54).

The hunters in *vangerengere*'s world used to go as far as means of transport could carry them and wherever it might be possible to find prey. Today they go as far as airplanes can carry them. A 2002 study showed that some 25 percent of Zimbabwean parents had been migrant workers of temporary duration in South Africa at some point in their lives (Tevera and Zinyama 2002). Today around three million Zimbabweans are estimated to be living outside Zimbabwe, dotted around the world, most of them in South Africa and the southern African region, and the rest in the English-speaking world (Crush and Tevera 2010). They might be far away but they are returning home in their numerous ways.

For all of us, as we left home, our mothers and fathers or kin came before *Mwari* to ask that we be spiritually guided in our mobilities into the world. Some used the way of Jesus Christ to get this message to *Mwari*; others went via their *midzimu*. Either way, we departed with a sense of risk taking that hunters adopt when going into the dark forests inhabited by other spirits. As was true for our fathers' sojourns in *valungu*'s world—in Triangle Sugar Estates, Joni, and Salisbury—today the guided ones are known by what they reap and send home. Read from the village, guided mobilities are still manifest in the family member based in the diaspora who sends money home to take care of her or his parents, her husband, his wife, their children, to buy a house in town, and build one at home. The lost ones are those whose hunt is perennial when all hunts must end, who have become of the diaspora, married foreign nationals, and who will not return home.

At issue is the question of what constitutes a return. Is it a matter of physically lifting oneself and traveling by plane back home? For the government, thus far the return, especially for skilled Zimbabweans, is a matter of physically lifting oneself and flying back to "build your country.

Your skills are required back home. The land awaits you; come and farm."
At the family level, returning physically when a close family member dies
is good, but the return that increasingly counts is the work that one is able
to do by being away from home: of sending money to bury a deceased
relative in an expensive casket, buy grain, send the children of a brother
or one's siblings to school, build parents a nice house, buy good clothes
for all. Common experiences reflect a sense of frustration with once-trusted
kin who now prey on the relative in the diaspora. Because the remittance
marks a disembodied return (that is, returning home through remittance),
too often because the sender cannot come due to the expense (the money
for airfares could be better used for something more tangible if sent in place
of the sender), a brother or sister, for example, and even parents use the
money for entirely different, often selfish purposes.

Today, Zimbabwe lies at a confluence of two realms of experience. The
first is that of its citizens who remained at home, endured the hard decade
from 2000 to 2008. Through their various mobilities, these citizens transformed the informal sector into the most dependable form and source of
livelihood. It says a lot that every time that the head of state wished to
travel, or when the government was paying a substantial installment in
debts owed to Mozambique and other countries for fuel and electricity
deliveries, the value of the Zimbabwe dollar vis-à-vis the rand, U.S. dollar,
and pound fell sharply.

We soon discovered that the government had come to the street like
everybody else and bought out all the foreign currency from the traders at
Roadport (Harare) and "The World Bank," downtown Bulawayo's parallel
foreign currency market nicknamed after the Bretton Woods institution. It
became easy for ordinary people in the streets to know that the government
had "raided" the market to get money to buy fuel or pay for VIP trips
abroad: the exchange rate shot through the roof. The slower the accumula-
tion of new forex through remittances from abroad, returning citizens, or
cross-border trading, the more Zimbabwe dollars we had to pay for a U.S.
dollar, British pound, South African rand, or Botswana pula. The weaker
the Zimbabwe dollar, the higher the prices for food—almost all of it now
imported from South Africa by "informal" (cross-border) traders. Inflation
was no longer a word that a finance minister or university economics guru
monopolized; everybody knew and lived its reality every day and itinerated
accordingly. The people who dealt with currency were usually women who
never went far in school, yet they were experts in calculating exchange
value. By contrast, the economic sanctions had resulted in the closure of
many foreign accounts, making it impossible for banks to access offshore

lines of credit, conduct international money transfers, and keep financial intermediation going (Nzaro, Njanike, and Munenerwa 2011).

This was also the time when a nephew who was not doing great in secondary school reminded his uncle when asked about the poor report: "*Vanhu vasina kuenda kuchikoro ndivo vari kutoita musoro mazuva ano*" (It's those who did not go to school who are having a good life nowadays). He was right. The salary was now worthless, with workers actually subsidizing their employers' wage bill by walking to work, staying in rural areas instead of the now out-of-reach city rentals. Many quit to go and deal in currency, or trekked to South Africa, sneaking without proper travel documents across the Limpopo River. Women became domestic workers, some became prostitutes; men became hawkers at traffic lights, some became robbers.

The nephew I just quoted seemed to be making a valid point: the problem with the Zimbabwean curriculum was that it trained people for formal employment. But who would give the graduates of the high schools, colleges, and universities a job when unemployment in the formal sector was over 80 percent? By contrast, the informal sector was growing because everything was now being sold on the "parallel market." Formal skills had no practical application—only informal ones, in an informal economy!

The point I have been trying to make is this: that hunting is only one of many transient workplaces that animate African creativities and life. Some of the spiritualities that anchored them may now be gone, the ancestors replaced by Jesus, and the tyranny of distance arrested by faster and longer-range means of communication and transport. However, the de- or re-spiritualized aspects of such creativities remain, sometimes in the original essence, other times in modified form. Excavating African creativities is going to take a systematic rethinking of methodologies. While discourses on Africa have tended to import concepts from outside to order the empirical evidence, this book was trying to suggest that a more nuanced approach lies in studying African thought and practice on its own terms. Such a method may be approached, for example, through close attention to mobility as an archive from which to uncover invisible and invisibilized knowledge of Africans. Such knowledge relates to cultures, philosophies, and practices of mobility as a workspace, a way of doing work, a transient workplace, and people as mobile bodies-at-work.

The hunt not only combines mobility and production to create transient workplaces. It also is an example of how a society can move from the villages—from clusters of physical homesteads bound into a community based on kinship ties—into the forest with the object of performing work on the move. The hunt as guided mobility, the construction of *hunza* or *goji* for hunting, the *nhimbe*, and *hupocha* affirm extensions of society into factory and society as a factory. Africa's still strong sense of community, best captured in the philosophy of *hunhu* or *ubuntu* (I am because we are), is no longer just social relations but a mobilities-centered mode of social innovation as well. In all these instances, the imperative to increase production or to enhance efficient execution of tasks necessitates the pooling of labor to generate mechanical power. These cultures of work are overlooked in studies that are concerned with the animal aesthetic, while hunting is reduced to "bushmeat," and "illegal hunting" is left unproblematized (Lindsey et al. 2011).

By calling for a rethinking of the hunt not as criminal behavior but a site of innovation that has now been rendered criminal, I am not trying to suggest a free-for-all in the national parks. Nor do I mean that all poachers have selfless intentions that derive from the need to survive; to the contrary I believe that there are hunting activities that benefit only a few individuals or rich syndicates. The issue for me is how to recover and utilize

Figure 9.1
For example, no chiefs or spirits of the forest in times past would have tolerated poisoning these two elephants and leaving the meat to rot.
Source: Edson Gandiwa, 2012

the precious knowledge, innovations, and practices that risk getting lost when hunting is criminalized and when those who engage in the practices are punished to the point not only of driving them underground but also forcing them to abandon these practices altogether. Seeking refuge in saying that "This is the twenty-first century and we cannot go back to the nineteenth century" would be a good option if the practices discussed in this book were no longer in use. That is not the case. Poaching continues even today, CAMPFIRE or no CAMPFIRE. The game reserve remains a fortress, the meaning of animals reduced to their aesthetic and $-equivalence. The concept and indeed current practice of "biodiversity" de-spiritualizes the relationship between local people, the forest, and animals.

The option of antagonizing entire communities by stripping them of their land to make way for game reserves, and then keeping them out at gunpoint or distancing them from the hunt through fences and fines has been in force for nearly a century. The projects that have been initiated from the tourism and conservation side have been concerned almost exclusively with saving the skin of the animals so that they continue to aesthetically entertain the rich and the outsiders. The game reserve can only survive in Africa if it embodies with a full heart the values and knowledge of ordinary people who are its neighbors. Such a decolonization of nature in Africa requires a paradigm shift that asks what nature Africans have historically known and practiced, and not simply to mobilize them as faithful employees. The entire ecological mindset governing the forest—and not just animals—requires new thinking if the Hwange National Park scenario is not to become a norm.

Notes

Introduction

1. Comment from the audience during my presentation of "Africa in Technology/ Technology in Africa," 24th Annual Nicholas Mullins Lecture, Department of Science and Technology in Society, Virginia Tech, February 22, 2013.

2. Comment from the audience during my "Usage Engineers as a Force behind the Movement of Technology: Part 1 Guns in Pre-colonial Africa," seminar paper, Department of Science and Technology Studies, Cornell University, September 10, 2013.

2 The Professoriate of the Hunt

1. These interviews are found in Simon Taylor's documentary *Finding Shadrek* (June 24, 2011), http://vimeo.com/25563235.

2. Ibid.

3. Interview with Solomon Bvekenya, Makuleke Cultural Centre & Homestay, November 5, 2010: Day 6. Black Bvekenyas Project (Hereafter BBP) File No. Mov050_1; Interview with Nyanisi Sarah Chauke, Boxahuku Village, July 21, 2008. TKAV Earthwatch/Widows and Elderly Women.

4. Ximuwu/Baobab, May 30, 2008. TKAV Earthwatch/Field Reports/Indigenous Biotechnology.

5. Trip to Pafuri Camp and Visit to Historical Sites in Kruger, June 27, 2008. TKAV Earthwatch/Field Reports/Joint Field Notes.

6. Nkuhlu/Natal Mahogany, May 30, 2008. TKAV Earthwatch/Field Reports/ Indigenous Biotechnology.

7. Mbhandhu/Mbholovisani or Apple/Rain Tree, May 30, 2008. TKAV Earthwatch/ Field Reports/Indigenous Biotechnology.

8. Ndhulwane or Poison Apple Plant, May 30, 2008. TKAV Earthwatch Expedition Field Reports/Indigenous Biotechnology.

9. Nkonono/Silver Cluster Leaf Tree, May 30, 2008. TKAV Earthwatch Expedition Field Reports/Indigenous Biotechnology.

10. Xanatsi/Mopane Tree, May 30, 2008. TKAV Earthwatch Expedition Field Reports/Indigenous Biotechnology.

11. Nsihani/Giant Raisin Shrub, June 20, 2008. TKAV Earthwatch Expedition Field Reports/Indigenous Biotechnology.

12. Ntoma/Jackal Berry Tree, June 20, 2008. TKAV Earthwatch Expedition Field Reports/Indigenous Biotechnology.

13. Ndhenga/Sickle Bush Creeper, June 20, 2008. TKAV Earthwatch Expedition Field Reports/Indigenous Biotechnology.

14. Interview with Hasani Chauke, Makuleke, January 11, 2011. TKAV/Co-Innovation Film Project Day 1/History of Fire.

15. Elmon Chauke Interviews Elias Nkuna (born 1943), Makuleke Village, December 11, 2011. TKAV Ecology Interviews/Hunters.

16. Interview with Hlengani Isaac Khosa (born August 1925) and Nwa France, Makuleke, June 23, 2011. TKAV Archives/Hunting.

17. Interview with Solomon Bvekenya, Makuleke, November 4, 2010. BBP/Mov04E_1.

18. Interview with Khosa and Nwa France. TKAV.

19. Interview with Nkuna. TKAV.

3   The Coming of the Gun

1. Interview with Solomon Bvekenya, Makuleke, November 5, 2010. BBP/Mov04F_1.

2. J. Fleetwood Churchill (Rhenosterpoort), September 13, 1856, 12. South African Historical Archives (SAHA), Transvaal Archives, File A17.

3. Native Commissioner Chibi to Superintendent of Natives, Victoria, March 6, 1912. NAZ NVC1/1/10.

4. Cpl. G. Pierce, C. 427/15, BSA Police, Chibi, to the District Superintendent, BSA Police, Victoria, February 5, 1916. NAZ A3/18/30/22.

5. NC Chibi to CNC Salisbury, March 14, 1898. NAZ N9/1/4.

6. District Boundaries: CNC to The Secretary Department of Administrator, September 6, 1920: "Native Department Station at Nuanetsi." NAZ N3/8/8.

7. Staff Officer, BSA Police, to Superintendent Victoria, February 28, 1923. NAZ S917/A312/481/A; Herbert Taylor, CNC, to Secretary of Native Affairs, November 29, 1923. NAZ S917/13/481; Lt-Col. J. Foulinsons, Assistant Commissioner BSA Police to Staff Officer Salisbury, July 6, 1924. NAZ S917/13/481.

8. Pierce to the District Superintendent. NAZ A3/18/30/22.

9. NC Chibi, Confidential Report to the CNC 9/10/14. NAZ N3/14/3; NC Ndanga, Report to CNC 30/9/14. NAZ N3/14/3; NC Chibi, Confidential Report to the CNC 9/10/14. NAZ N3/14/3.

10. Interview with Solomon Bvekenya, on Nthlaveni Block H Hill, November 4, 2010. BBP/Mov04C_1 11//2010.

11. Interview with Bvekenya, November 4, 2010. BBP/Mov04C_1 11//2010.

12. Various reports. 305/P5 Poaching Records, 1968–1972, 3. Mabalauta Field Station Archives.

13. Interview with Bvekenya, November 4, 2010. BBP/Mov04C_1 11//2010.

14. Ibid.

4 Tsetse Invasions

1. Secretary, Salisbury, to Mining Commissioner, Lomagundi, September 6, 1894. NAZ M 1/1/1; Resident Magistrate, Umtali, to Public Prosecutor, Salisbury, October 23, 1894. NAZ J 1/1/1.

2. E. G. Kay (Glossinologist), "Report on the West Sebungwe Anti-Tsetse Spraying Operation July–October 1967," 1. SACEMA/TA.

3. T. Lees May (Director of Veterinary Services), "The Land Settlement Problem in Rhodesia," no date but post-1965, 1. SACEMA/TA.

4. Ibid.

5. Graham Child (Director, Department of National Parks and Wild Life Management) and Thane Riney (Australian National Parks and Wildlife Service), "An Examination of the Returns from Tsetse Control Hunting in Zimbabwe in 1919 through 1958," manuscript written in 1980, 1. SACEMA/TA.

6. J. K. Chorley, "The Tsetse Fly Problem in Southern Rhodesia," March 17, 1937, 10. SACEMA/TA.

7. D. F. Lovemore, "The Present Tsetse-Trypanosomiasis Situation: A Summary," February 23, 1968, 1. SACEMA/TA.

8. "Notes Taken at a Meeting Held on the 1st February, 1962, to Discuss Tsetse and Trypanosomiasis in Relation to Development Projects in Mashonaland East," 1. SACEMA/TA.

9. T. J. Casewell, "The Control of Tsetse and Trypanosomiasis in the North Eastern Districts of Rhodesia," Branch of Tsetse and Trypanosomiasis Control, Department of Veterinary Services Rhodesia, March 22, 1968, 6. SACEMA/TA.

10. Director, Tsetse Fly Operations, to Secretary Trypanosomiasis Commission, "Land Settlement and Tsetse Reclamation," April 7, 1953, 16. SACEMA/TA.

11. Casewell, "Control of Tsetse and Trypanosomiasis," 1–3.

12. Ibid., 6.

13. John Ford, "The Advance of the *Glossina morsitans* and *Glossina pallidipes* into the Sabi and Lundi River Basins, Southern Rhodesia," no date, 1. SACEMA/TA.

14. W. P. Boyt (Chief Veterinary Officer [Trypanosomiasis]), "Trypanosomiasis Southern Rhodesia 1964: A Review," February 4, 1964, 26. SACEMA/TA.

15. Director, Tsetse Fly Operations, "Land Settlement and Tsetse Reclamation," 1.

16. Ibid.; Boyt, "Trypanosomiasis Southern Rhodesia 1964," 26.

17. Ford "Advance of the Glossina," 1.

18. Boyt, "Trypanosomiasis Southern Rhodesia 1964," 28.

19. J. A. Farrell (Entomologist), "A Preliminary Report on the Humani and Devuli Spraying Operation of August–September, 1960," 3. SACEMA/TA.

20. Ibid.

21. Ibid.

22. J. K. Chorley, Director Tsetse Fly Operations, to the Secretary to the Federal Ministry of Agriculture, "Trypanosomiasis Threat: Ndanga Ranches," July 26, 1955, 1. SACEMA/TA.

23. Ibid., 2.

24. Ibid.

25. Ibid., 3.

26. Farrell, "Preliminary Report," 5–6.

27. Ibid.; Boyt, "Trypanosomiasis Southern Rhodesia 1964," 28.

28. Chorley, "Trypanosomiasis Threat," 1.

29. J. A. Farrell (Entomologist), "The Vegetation of the Sabi East Bank Area, Chipinga District," November 1960. SACEMA/TA.

30. Farrell, "Preliminary Report," 3.

31. Boyt, "Trypanosomiasis Southern Rhodesia 1964," 29.

32. Ibid., 31.

33. Ibid.

34. Ibid.

35. Ford, "Advance of the Glossina," 2.

36. Ibid., 1.

37. K. E. W. Boyd, "Tsetse Fly Survey of the Lower Sabi/Lundi Area, 1954." NAZ S3106/11/1/8 Sabi Valley 1953–1955.

38. Ford, "Advance of the Glossina," 1.

39. Boyd, "Tsetse Fly Survey."

40. Boyt, "Trypanosomiasis Southern Rhodesia 1964," 32.

41. Boyd, "Tsetse Fly Survey."

42. Ford, "Advance of the Glossina," 2.

43. Ibid.

44. R. Goodier, "A Survey of the Trypanosomiasis Situation among Donkeys in the Lower Sabi Area (East Bank) 1956–1957," 1–5. SACEMA/TA.

45. Ford, "Advance of the Glossina," 1.

46. Boyt, "Trypanosomiasis Southern Rhodesia 1964," 33.

47. Ibid.

48. NAZ, S1194/1645/3/1: Chief Entomologist to Senior Forest Officer, Oct. 28, 1932; Secretary Department of Agriculture and Lands to the Minister, 1645/155.

49. The entire file NAZ, S1193/T4/9 Tsetse Fly Melsetter Border Area 1925–1929.

50. Files NAZ, S1215/1880/1 Nuanetsi Ranch—Boundary Fencing 1935–1943; S3106/11/1/3 Sabi Valley 1940–1944; S2136/58779/66 Fencing Prohibited Area, South Eastern Districts 1937.

51. Boyd, "Tsetse Fly Survey."

52. Ford, "Advance of the Glossina," 2–3.

53. Director Tsetse Fly Operations to K. E. W. Boyd, Entomologist, April 20, 1954. NAZ S3106/11/1/8 Sabi Valley 1953–1955.

54. Ford, "Advance of the Glossina," 1.

55. Director, Tsetse Fly Operations, "Land Settlement and Tsetse Reclamation," 1.

56. Ford, "Advance of the Glossina," 2.

57. Ford, "Advance of the Glossina," 3.

58. Interview with Jerry Masevhe, nephew of Matahe Dodorolo Kufakuzvida (age unknown but believed to be over ninety years old), Chibwedziva, December 4, 2010. BBP/Mov092 and Mov093.

59. Director Tsetse Fly Operations to K. E. W. Boyd.

60. Interview with Timothy Sumbani, Chipinda Pools, December 14, 2010. BBP/MovOBF.

61. Interview with Jerry Masevhe. BBP/Mov092 and Mov093.

62. Farrell, "Preliminary Report," 5–6.

63. Boyt, "Trypanosomiasis Southern Rhodesia 1964," 34.

64. Ford, "Advance of the Glossina," 3.

65. Ibid., 4.

66. Ibid., 3.

67. Ibid.

68. J. A. K. Farrell, "Insecticide Spraying Operations, Chiredzi River, September–November 1959," November 21, 1959, 1. SACEMA/TA.

69. Boyt, "Trypanosomiasis Southern Rhodesia 1964," 36.

70. Ibid.

71. Ibid., 37.

72. Ibid., 38.

73. Ford, "Advance of the Glossina," 3.

74. Farrell, "Control of a Tsetse Fly (Glossina) Advance," 2.

5   The Professoriate of the Hunt and the Tsetse Fly

1. "Report of the Assistant Director of Veterinary Services (Tsetse and Trypanosomiasis Control) for the Year Ending 30 September 1961," 217. SACEMA/TA.

2. Annual Report of the Government Entomologist, 1919. SACEMA/TA.

3. Annual Report of the Chief Entomologist, 1920, 2. SACEMA/TA.

4. Annual Report of the Chief Entomologist, 1921. SACEMA/TA.

5. Annual Report of the Chief Entomologist, 1923, 2. SACEMA/TA.

6. "Report of the Director of Tsetse and Trypanosomiasis Operations and Reclamation for the Year Ended 30 September 1958," 202. SACEMA/TA.

7. "Report of the Director of Tsetse and Trypanosomiasis Operations and Reclamation for the Year Ended 30 September 1960," 225. SACEMA/TA.

8. "Report of the Director 1960," 227. SACEMA/TA.

9. "Report of the Director 1960," 230. SACEMA/TA.

10. "Report of the Director 1960," 231. SACEMA/TA.

11. "Report of the Assistant Director of Veterinary Services (Tsetse and Trypanosomiasis Control) for the Year Ending 30 September 1961," 217. SACEMA/TA.

12. J. Mackinnon, "Report of the Director of Veterinary Services for the Year Ending 30 September 1962," 205. SACEMA/TA.

13. G. F. Cockbill, "Report of the Assistant Director of Veterinary Services (Tsetse and Trypanosomiasis Control) for the Year Ending 30 September 1964," 2. SACEMA/TA.

14. Ibid., 1.

15. D. F. Lovemore, "Report of the Assistant Director of Veterinary Services (Tsetse and Trypanosomiasis Control Branch) for the Year Ending 30 September 1967," 2. SACEMA/TA.

16. J. K. Chorley, "Tsetse Fly Operations: Short Survey of the Operations by Districts for Year Ending 31st December 1944," 7. SACEMA/TA.

17. Ibid., 12.

18. "Report of the Acting Director of the Department of Tsetse and Trypanosomiasis Operations and Reclamation for the Year Ended 30 September 1956," 148. SACEMA/TA.

19. "Report of the Acting Director of the Department of Tsetse and Trypanosomiasis Operations and Reclamation for the Year Ended 30 September 1957," 8. SACEMA/TA.

20. "Report of the Director of Tsetse and Trypanosomiasis Operations and Reclamation for the Year Ended 30 September 1958," 211. SACEMA/TA.

21. G. F. Cockbill, "Report of the Assistant Director of Veterinary Services (Tsetse and Trypanosomiasis Control) for the Year Ending 30 September 1965," 30. SACEMA/TA.

22. Lovemore, "Report of the Assistant Director," 2–3.

23. G. F. Cockbill, "Annual Report of the Branch of Tsetse and Trypanosomiasis Control for the Year Ending 30 September 1968," 2. SACEMA/TA.

24. Ibid., 19, 21.

25. Ibid., 25.

26. G. F. Cockbill, "Annual Report of the Branch of Tsetse and Trypanosomiasis Control for the Year Ending 30 September 1970," 21. SACEMA/TA.

27. J. K. Chorley, Entomologist, Agricultural Laboratory, Department of Agriculture, Salisbury, to The Staff Officer, BSA Police, Salisbury, October 24, 1934: "Return of MH Rifle Held at Police Station, Gatooma." NAZ S3099/15 Martini-Henry Rifles for Tsetse Fly Operations 1933–1940.

28. J. K. C., Acting Chief Entomologist, Agricultural Laboratory, Salisbury, A. Aitken, Esq., Ranger i/c Tsetse Fly Operations, Gatooma, December 21, 1938: "Supply of Ejectors." NAZ S3099/15 Martini-Henry Rifles for Tsetse Fly Operations 1933–1940.

29. Chief Entomologist to Mr. C. Bailie-Barry, Assistant Tsetse Fly Ranger, c/o J. J. Struthers, Esq., Tsetse Fly Operations Gatooma, July 9, 1936: "Martini-Henry Rifle." NAZ S3099/15 Martini-Henry Rifles for Tsetse Fly Operations 1933–1940.

30. Rupert W. Jack, "Annual Report of the Division of Entomology for Year Ending 31st December 1937." SACEMA/TA.

31. J. K. Chorley, Entomologist, Agricultural Laboratory, Department of Agriculture, Salisbury, to A. Aitken, Gatooma, C. Beattie, Darwin, Bond, Sinoia, J. Hall, Miami, J. B. M. Powell, Sinoia, J. J. Struthers, Bulawayo, October 16, 1937: "Conservation of Martini-Henry Ammunition." NAZ S3099/15 Martini-Henry Rifles for Tsetse Fly Operations 1933–1940.

32. Ibid.

33. J. K. Chorley, "Report of the Division of Entomology for the Year Ending 31st December 1938," 9. SACEMA/TA.

34. Ibid.

35. J. K. Chorley, "Tsetse Fly Operations: Short Survey of the Operations by Districts for Year Ending 31st December 1940," 3. SACEMA/TA.

36. J. K. Chorley, "Tsetse Fly Operations: Short Survey of the Operations by Districts for Year Ending 31st December 1941," 3. SACEMA/TA.

37. J. K. Chorley, "Tsetse Fly Operations: Short Survey of the Operations by Districts for Year Ending 31st December 1944," 4. SACEMA/TA.

38. J. K. Chorley, "Tsetse Fly Operations: Short Survey of the Operations by Districts for Year Ending 31st December 1945," 8. SACEMA/TA.

39. J. K. Chorley, Entomologist, to the Chief Entomologist, Agricultural Laboratory, Salisbury, October 18, 1937: "Replacement of Martini-Henry Rifles: 'Tsetse Fly Operations.'" NAZ S3099/15 Martini-Henry Rifles for Tsetse Fly Operations 1933–1940.

40. Ibid.

41. J. A. Whellan (Entomologist, Division of Agriculture and Lands), "Tsetse Fly in Southern Rhodesia 1950–51," November-December 1953, 3. SACEMA/TA.

42 J. K. Chorley, "Tsetse Fly Operations: Short Survey of the Operations by Districts for Year Ending 31st December 1946," 4. SACEMA/TA.

43. Ibid.

44. J. K. Chorley, "Report of the Division of Entomology for the Year Ending 31st December 1947," 11–12. SACEMA/TA.

45. Ibid., 11.

46. Ibid., 12.

47. "A Review of the Tsetse Situation in S. Rhodesia, 1948," 4–5. SACEMA/TA.

48. "Tsetse Fly in S. Rhodesia, 1949, September–October 1950," 6. SACEMA/TA.

49. J. A. Whellan, "A Review of the Tsetse Fly Situation in S. Rhodesia, 1948," 5. SACEMA/TA.

50. J. A. Whellan, "Tsetse in S. Rhodesia, 1949," 9. SACEMA/TA.

51. J. K. Chorley, "Annual Report of the Director of Tsetse Fly Operations, 1951–2," 4. SACEMA/TA.

52. J. K. Chorley, "Annual Report of the Director of Tsetse Fly Operations for the Year Ending 30 September 1953," 4. SACEMA/TA.

53. Ibid., 5.

54. J. K. Chorley, "Annual Report of the Director of Tsetse Fly Operations for the Year Ending 30 September 1954," 2. SACEMA/TA.

55. Chorley, "Annual Report 1954," 4. SACEMA/TA.

56. "Report of the Acting Director of the Department of Tsetse and Trypanosomiasis Operations and Reclamation for the Year Ended 30 September 1956," 151. SACEMA/TA.

57. "Report of the Director of Tsetse and Trypanosomiasis Operations and Reclamation for the Year Ended 30 September 1959," 197. SACEMA/TA.

58. Graham Child and Thane Riney, "An Examination of the Returns from Tsetse Control Hunting in Zimbabwe in 1919 through 1958," manuscript written in 1980, 19. SACEMA/TA.

59. "Report of the Director of Tsetse and Trypanosomiasis Operations and Reclamation for the Year Ended 30 September 1960," 228. SACEMA/TA.

60. Ibid., 230.

61. Ibid., 233.

62. Ibid., 234.

63. Ibid.

64. Chorley, "Annual Report 1953," 7.

65. Interview with Gazeni Komundela, Chibwedziva, December 11, 2010. BBP/Mov0B4.mov.

66. Interview with Hanyani Chauke, Sabhuku Bhaule, Area Bhaule, January 10, 2011. BBP/Mov0A6.mov.

67. Ibid.; interview with Sabhuku Joshua Hasani Mahuhushe, Chibwedziva, Sala Village, December 11, 2010. BBP/Mov0B3.mov.

68. Interview with Gazeni Komundela.

69. Ibid.

70. Ibid.; interview with Mavuyani Koteni Sumbani, December 14, 2010. BBP/Mov0C0.mov.

71. Interview with Solomon Bvekenya, November 5, 2010. BBP/Mov0S0_1 11/5/2010.

72. Camera Sweep Showing the Layout of Chibwedziva Business Center, January 1, 2011. BBP/Mov0B2.mov.

73. Whellan, "Tsetse in S. Rhodesia, 1949," 6.

74. Senior Entomologist, for Director Tsetse Fly Operations, Causeway, to the Game Officer, May 27, 1953: "Chiredzi ICA Committee." NAZ S3106/11/1/7 Sabi Valley 1952–1953.

75. Interview with Solomon Bvekenya, November 4, 2010. BBP/Mov04C_1. 11//2010.

76. Interview with Piet Barnard, Mapikule Village, December 6, 2010. Mov10D; interview with Lamson Watson Machiukele (Masivamele Royal Family, born 1933), Chibwedziva, December 6, 2010. BBP/Mov09C.

77. Interview with Solomon Bvekenya, November 5, 2010. BBP/Mov04F_1 11/5/2010.

78. Ibid.

79. Interview with Solomon Bvekenya, November 5, 2010. BBP/Mov04F_1 11/5/2010.

80. Interview with Piet Barnard, Mapikule Village, December 6, 2010.

81. Child and Riney, "An Examination," 19.

82. Interview with Piet Barnard, Mapikule Village, December 6, 2010.

83. J. K. Chorley, "Annual Report of the Director of Tsetse Fly Operations for the Year Ending 30 September 1955," 12. SACEMA/TA.

84. R. Goodier, "A Survey of the Tsetse and Game Position in the Chirundu Area of the Zambesi Valley," August 1956, 4. SACEMA/TA.

85. Child and Riney, "An Examination," 19.

86. Whellan, "Tsetse in S. Rhodesia, 1949," 9.

87. J. K. Chorley, "Annual Report of the Director of Tsetse Fly Operations for the Year Ending 30 September 1955," 6. SACEMA/TA; interview with Mugocha Julius Mavasa, January 1, 2011. BBP/Mov0F3.mov.

88. Whellan, "Tsetse Fly in Southern Rhodesia 1950–51," 3. SACEMA/TA.

89. "Chiredzi Intensive Conservation Area: Minutes of a Continued Annual General Meeting of the Chiredzi ICA Held at Messrs Earl and Huntley, Chiredzi Ranch, Wednesday 29th April 1953, 10am." NAZ S3106/11/1/7 Sabi Valley 1952–1953.

90. Goodier, "Survey of the Tsetse and Game Position in the Chirundu Area," 5.

91. B. O. E. Cooke-Yarborough, Tsetse Fly Operations, P. O. Craigmore, to the Chief Entomologist, Box 387 Salisbury, 4/8/45: "Mileage for Private Car." NAZ S3106/11/1/6 Sabi Valley 1944–1946.

92. J. K. C., Chief Entomologist, to Secretary Department of Agriculture and Lands, 15th August 1945: Motor Mileage Allowance: Tsetse Fly Rangers." NAZ S3106/11/1/6 Sabi Valley 1944–1946.

93. Senior Entomologist, for Director Tsetse Fly Operations, Causeway, to the Game Officer, 27th May 1953: "Chiredzi ICA Committee." NAZ S3106/11/1/7 Sabi Valley 1952–1953.

94. G. Davison, "Review of Hunting Operations, June 1971–March 1972," May 4, 1972, 6. SACEMA/TA.

95. D. F. Lovemore, "Report by Mr. Davison, Glossinologist I/C Coordination of Hunting Operations Entitled 'Review of the Hunting Operations over the Past Season,'" July 14, 1971, 1. SACEMA/TA.

96. Ibid.

97. Ibid.

98. Davison, "Review of Hunting Operations," 1.

99. Ibid., 5.

100. Child and Riney, "An Examination," 15.

101. Ibid., 18.

102. R. Heathcote to Assistant Director of Veterinary Services (Tsetse and Trypanosomiasis Control), "A Review of Hunter Operations," January 26, 1972, 1. SACEMA/TA.

103. J. K. Chorley, Director Tsetse Fly Operations, to the Secretary to the Federal Ministry of Agriculture, "Trypanosomiasis Threat: Ndanga Ranches," July 26, 1955. SACEMA/TA.

104. Davison, "Review of the Hunting Operations," 3.

105. Interview with Mugocha Julius Mavasa, January 1, 2011. BBP/Mov0F3.mov.

106. Davison, "Review of the Hunting Operations," 3.

107. Ibid., 4–5.

6   Poaching as Criminalized Innovation

1. Interview with Lamson Watson Machiukele (Masivamele Royal Family, born 1933) in Chibwedziva, December 6, 2010. BBP/Mov09C.

2. Interview with Solomon Bvekenya, November 4, 2010. BBP/Mov04C_1. 11//2010.

3. Interview with Hanyani Chauke, Sabhuku Bhaule, Area Bhaule, January 10, 2011. BBP/Mov0A6.mov.

4. Chief Native Commissioner's Office, Salisbury, to the Secretary of Administrator, January 24, 1917: "Cattle Theft by German (Diegal) Living in Portuguese Territory": Copy of affidavit signed by Gaba before P. Forrestall, NC Chibi, December 21, 1916. NAZ A3/18/20/30/22 Recruiting Illicit 1915–1918.

5. Interview with Hanyani Chauke; interview with Elizabeth Mazichina, Mahenye, January 5, 2011. BBP/Mov10D.mov; interview with Nwa Jeke Shimbani Njakanjaka, December 2, 2010. BBP/Mov086.mov.

6. Interview with Gazeni Komundela, December 11, 2010. BBP/Mov0B4.mov.

7. Ibid.

8. Interview with Mavuyani Koteni Sumbani, December 14, 2010. BBP/Mov0C0.mov.

9. Ibid.

10. Ibid.

11. "Cattle Theft by German (Diegal)": Copy of affidavit.

12. Interview with Mugocha Julius Mavasa, January 1, 2011 BBP/Mov0F3.mov.

13. Ibid.

14. Mubhulachi Mavasa's son Mugocha Julius Mavasa confirms what Bvekenya told his biographer Thomas Bulpin. Interview with Mugocha Julius Mavasa.

15. Ibid.

16. Ibid.

17. Interview with Lamson Watson Machiukele.

18. Interview with Khosa and Nwa France. TKAV Archives: Hunting.

19. Ibid.

20. CNC to Minister of Commerce; Assistant NC Bawden to Supt. of Natives, Fort Victoria. NAZ S914/12.

21. Bullock, CNC to Secretary to Premier, November 7, 1932: "Suggested Proclamation of Game Reserve: Chipinda Pools Area." NAZ S1194/1645/3/1; D. Townley to Gilchrist, October 5, 1934; S. Rogers Divisional Road Engineer to Chief Road Engineer, June 12, 1934. S914/12/1B.

22. Kelly Edwards to Acting Secretary of Agriculture, April 8, 1933: "Game Reserves." NAZ S1194/1645/3/1.

23. Acting Secretary of Commerce and Transport to Col. Deneys Reitz, Minister of Lands, Pretoria, September 28, 1934: "Gonarezhou Game Reserve." NAZ S914/12/1B; Ministrie van Lande, Pretoria, October 5 and 15, 1934. NAZ S914/12/1B.

24. Supt. Southern Rhodesia Publicity Bureau, Bulawayo to Secretary Agriculture, May 9, 1934; Secretary of Commerce to Minister of Agriculture, 29.3.34. NAZ S914/12/1B.

25. Chipinda Pools—Proposed Reserve 1933–1934, F. Hackney, Midlands Hotel, Gwelo, to the Secretary of Commerce, February 19, 1934. NAZ S914/12/1D.

26. Secretary of Commerce to Hackney, February 20, 1934; Secretary of Commerce to Chief Clerk of Lands; the Minister 9/4/34; Undersecretary of Lands to Hackney, May 1, 1934. NAZ S914/12/1D; also Secretary of Agriculture to Secretary of Commerce, April 13, 1934: "Controlled Shooting." S914/12/1B.

27. C. Ashley-Belbin to Minister of Agriculture, May 7, 1935. NAZ S1194/1645/3/1.

28. Ibid.; H. G. Mundy, Secretary of Agriculture and Lands to Ashley-Belbin, May 29, 1935; CNC to Secretary to the Prime Minister, November 25, 1933; Chief Forest Officer to Secretary of Agriculture, May 16, 1935. NAZ S1194/1645/3/1.

29. Manager Nuanetsi Ranch to Magistrate Gwanda, October 31, 1932; Secretary of Agriculture to CNC, November 3, 1932. NAZ S1194/3/1; the Assistant NC Nuanetsi, J. Bawden, to Supt. of Natives, Fort Victoria. NAZ S914/12.

7 *Chimurenga*: The Transient Workspace of Self-Liberation

1. Interview with Phineas Ngobeni, Makuleke, June 6, 2011. TKAV/Indigenous Ecology: Hunting.

2. Interview with Hanyani Hlayisi (born 1936), December 14, 2010. BBP/Mov0C0.mov.

3. Interview with Lamson Watson Machiukele (Masivamele Royal Family, born 1933), Chibwedziva, December 6, 2010. BBP/Mov09C.

4. Report of the Director of National Parks and Wild Life Management 1964, 24. NAZ SRG/3.

5. Report of the Director of National Parks and Wild Life Management Rhodesia 1968, 23. NAZ SRG/3.

6. Report of the Director of National Parks and Wild Life Management Rhodesia 1969, 11. NAZ SRG/3.

7. Interview with Bvekenya, November 4, 2010. Mov04C_1 11//2010.

8. Delineation Report Nuanetsi District: The Ngwenyenye or Marumbini Headmanship and Community, Chief Chitanga, Gonarezhou National Land, B. P. Kaschula, Ministry of Internal Affairs, Salisbury, June–August 1967. NAZ S2929/8/4; Report of the Director of National Parks and Wild Life Management Rhodesia 1967, 21. NAZ SRG/3.

9. Report of the Director of National Parks and Wild Life Management Rhodesia 1968, 23.

10. Interview with Bvekenya, October 31, 2010. BBP/Mov047–9 10/31/2010.

11. Interview with Bvekenya, November 5, 2010. BBP/Mov050_1 11/5/2010.

12. Interview with Phineas Ngobeni.

13. Interview with Bvekenya, November 5, 2010.

14. Ibid.

15. Ibid.; Wright 1972: 368. Interview with Phineas Ngobeni.

16. Interview with Bvekenya, November 5, 2010.

17. Ibid.

18. Ibid.

19. Ibid.

20. Ibid.

21. Interview with Bvekenya, November 4, 2010.

22. Ibid.

23. Interview with Timothy Sumbani, December 14, 2010. BBP/Mov0BF.mov.

24. Ibid.

25. Interview with Bvekenya, November 4, 2010.

26. Interview with Risimati Mapikule and Other Elders Teaching and Learning *Dhema* through Doing, December 14, 2010. BBP/Mov0CF.mov.

27. Interview with Bvekenya, November 5, 2010.

28. Ibid.

29. Ibid.

30. Ibid.

31. Ibid.

32. Camera Sweep Showing Video Clip of Masivamele Old School Group, December 2, 2010. BBP/Mov0A1.

33. Interview with Bvekenya, November 4, 2010.

34. MFS Archives 305/P5 Poaching Records, 1968–1972, 1.

35. Interview with Bvekenya, November 4, 2010.

36. Interview with Isaac "Roza" Khosa and Maria Hlungwani, June 23, 2011. TKAV; Interview with Hlengani Isaac Khosa, Makuleke, June 23, 2011. TKAV.

37. 305/P5 Poaching Records, 1968–1972. MFS Archives.

38. Ibid.

39. Ibid.

40. Ibid.

41. Ibid.

42. Ibid.

43. Ibid.

44. Ibid.

45. Interview with Bvekenya, November 5, 2010.

46. Ibid.

47. Ibid.; Interview with Piet Barnard.

48. Interview with Piet Barnard.

49. Ibid.

50. Interview with Annie Ngobeni and Ester Ngobeni, September 16, 2011. TKAV; Interview with Phineas Ngobeni; Interview with Piet Barnard.

51. Interview with Bvekenya, November 5, 2010.

52. Interview with Phineas Ngobeni.

53. Interview with Johannes "MuKorea" Sibanda, ex-ZIPRA guerrilla, January 1, 2011. BBP/Mov07B.mov.

54. Interview with Lamson Watson Machiukele.

55. Interview with Piet Barnard.

56. Ibid.

57. Interview with Annie Ngobeni and Ester Ngobeni, September 16, 2011. TKAV.

58. Interview with Phineas Ngobeni; Interview with Piet Barnard.

59. Ibid.

60. Interview with Annie Ngobeni and Ester Ngobeni.

61. Ibid.

62. Interview with Lamson Watson Machiukele.

63. Ibid.

64. Ibid.

65. Interview with Annie Ngobeni and Ester Ngobeni.

66. Interview with Lamson Watson Machiukele.

67. Ibid.

68. Ibid.

8 The Professoriate of the Hunt and International Ivory Poaching

1. 305/P-5/ Poaching Records, 1968–1972. MFS/305.

2. Ibid.

3. *Finding Shadrek.*

4. Ibid.

5. R. L. Murray, Warden Mabalauta to Chief Warden: "Juveniles: Poaching—Gonarezhou National Park," August 5, 1983. MFS/305/8/83.

6. Warden Palmer to ZRP Mpakati—Mabalauta Field Station, Gonarezhou National Park, February 29, 1984. MFS/305.

7. Westrop to OC ZRP Chiredzi, att. Supt. Chingosho, March 6, 1986: "Release of Juvenile Poachers—Police Mwenezi." MFS/305/3/86.

8. Ibid.

9. Ibid.

10. Interview with Mugocha Julius Mavasa, January 1, 2011. BBP/Mov0F3.mov.

11. Interview with Risimati Mbanyele, December 29, 2010. BBP/Mov0EE.mov.

12. Interview with Mugocha Julius Mavasa.

13. Warden to Director DPWLM, December 11, 1984. MFS/305.

14. D/136/89 A. W. J. Wood, for Director, to Warden Mabalauta, January 10, 1985: Armed Poaching in the Gonarezhou. MFS/305.

15. W. T. Takura, Chief Supt., Staff Officer (Legal Services) to the Commissioner of Police, to Secretary for Home Affairs. Attention: Mr. Tsomondo, The Director of Parks and Wildlife, March 11, 1985: "Powers of Arrest: Parks and Wildlife Authorities." MFS/305.

16. C. A. M'pamhanga for Acting Director, to All Staff, All Stations, April 17, 1985: Powers of Arrest/Use of Firearms—Parks and Wildlife Personnel. MFS/305/A/977.

17. M. R. Drury for Director to All Stations, November 8, 1985 (see Prime Minister Mugabe's warning to poachers, *The Herald*, November 5, 1985). MFS/305/A/934.

18. Interview with Solomon Bevkenya, November 5, 2010. BBP/Mov 050_1 11/5/2010.

19. Ref: Parks and Wildlife Offenses: Prosecution: "Juveniles." MFS/305.

20. P. G. E. Westrop, Senior Ranger to OiC ZRP Chiredzi: Att. Inspector Ndhlovu, October 31, 1985: Armed Poaching Cases—Gonarezhou National Park. MFS/305/10/85.

21. Ibid.

22. Ibid.

23. Westrop to OC ZRP Chiredzi, att. Supt. Chingosho, March 6, 1986: "Release of Juvenile Poachers—Police Mwenezi." MFS/305/3/86.

24. Ibid.

25. Ibid.

26. Ibid.

27. Ibid.

28. Interview with Risimati Mbanyele.

29. G. M. Nott, for Director to All Stations, November 1, 1983: "Evidence in Aggravation—Poaching Cases." MFS/05/A845/1/LEG.

30. "Parks and Wildlife Act—Particulars of Accused," no date. MFS/305/0262.

31. *Finding Shadrek.*

32. Ibid.

33. Ibid.

34. Ibid.

35. Ibid.

36. Ibid.

37. Ibid.

38. Ibid.

Conclusions

1. Interview with Koteni Sumbani, December 14, 2010. BBP/Mov0BF.mov.

References

Unpublished Primary Documents

State-generated Archives
National Archives of Zimbabwe (NAZ).

South African Centre for Epidemiological Modelling and Analysis (SACEMA).

South African Historical Archives (SAHA), Transvaal Archives.

Mabalauta Field Station (MFS), Gonarezhou National Park.

Field Interviews and Ethnography
The ethnographic material for this book is extracted from two projects I have conducted since 2008 that are dedicated to gathering (using digital cameras, camcorder, and computers) what ordinary people in the rural African countryside know and storing these resources within the communities for use in innovation. Brief histories of these projects are given below to orientate the reader on the nature of the archives, a fraction of which has been used in this book.

Traditional Knowledge of African Villages (TKAV), Makuleke, South Africa
TKAV emerged in 2008 as an initiative to collect and record the indigenous knowledge of rural communities to address our day-to-day challenges. Thus far we have produced more than 1 terabyte of video and audio materials covering topics such as indigenous energy strategies, environment and ecology, medicinal plants, folklore, music, art, proverbs, as well as the history, culture, and economy of the Makuleke dating back to circa 1750. I oversee the project while Elmon Magezi Chauke, a member of the Makuleke community, conducts the day-to-day research activities.

The Black Bvekenyas Project (BBP), Chibwedziva, Zimbabwe

The Black Bvekenyas Project began in 2010. Its purpose is to document the life and afterlives of the famous ivory poacher Cecil Barnard, whom locals called *Bvekenya*, from the perspectives of his black children and grandchildren. Through it, his black grandchildren are trying to tell not only the stories of their lives, but also the histories of maTshangana. Thus the project has a 20 GB store of knowledge already collected since 2010. Solomon Bvekenya, a grandson of the famous "poacher," is coordinating the interviews on a day-to-day basis, while I review material collected, identify gaps, and draft questions and potential leads for further interviews and site visits to ensure that the narratives are told as fully as possible.

Published Primary Sources

Indigenous Traditions: Poetry, Registers, Storytelling, and Music

Chimera, L. 1983. "Mbuya Chikonamombe." In *Nehanda Nyakasikana: Nhorido Dzokunyikadzimu*, comp. Ticha Jongwe, 174–175. Gweru: Mambo Press.

Chitsungo, M. 1983. "Vene Vemasango." In *Nehanda Nyakasikana: Nhorido Dzokunyikadzimu*, comp. Ticha Jongwe, 165–166. Gweru: Mambo Press.

Denhere, D. 1983. "Mwana Oenda Kune Imwe Nyika." In *Nehanda Nyakasikana: Nhorido Dzokunyikadzimu*, comp. Ticha Jongwe, 176–177. Gweru: Mambo Press.

Gowera, E. 1983. "Titungamirei." In *Nehanda Nyakasikana: Nhorido Dzokunyikadzimu*, comp. Ticha Jongwe, 161. Gweru: Mambo Press.

Greaves, Nick. 1996. *When Elephant Was King and Other Elephant Tales from Africa*. Cape Town: Struik.

Gwete, C. 1983. "Parwendo." In *Nehanda Nyakasikana: Nhorido Dzokunyikadzimu*, comp. Ticha Jongwe, 191–192. Gweru: Mambo Press.

Hodza, Aaron, compiler. 1979. *Shona Praise Poetry*. Oxford: Clarendon Press.

Hodza, Aaron C. 1980. *Denhe Renduri neNhorimbo*. Harare: Mercury Press.

Hodza, Aaron C. 1984. *Shona Registers*, vol. 1. Harare: Mercury Press.

Hove, C. 1983. "Musinauta Save." In *Nehanda Nyakasikana: Nhorido Dzokunyikadzimu*, comp. Ticha Jongwe, 112–113. Gweru: Mambo Press.

Hwiridza, D. 1983. "Auye Aronge Matare." In *Nehanda Nyakasikana: Nhorido Dzokunyikadzimu*, comp. Ticha Jongwe, 113–114. Gweru: Mambo Press.

Jongwe, Ticha. 1983. *Nehanda Nyakasikana: Nhorido Dzokunyikadzimu*. Gweru: Mambo Press.

Karambwe, W. K. 1983. "Kusuma Vadzimu Nezverwendo." In *Nehanda Nyakasikana: Nhorido Dzokunyikadzimu*, comp. Ticha Jongwe, 188–191. Gweru: Mambo Press.

Kunodziya, O. K. 1983. "Muzukuru Wopinda Mumasango." In *Nehanda Nyakasikana: Nhorido Dzokunyikadzimu*, comp. Ticha Jongwe, 184–185. Gweru: Mambo Press.

Madombi, F. 1983a. "Zvabva Mamuri Mashave." In *Nehanda Nyakasikana: Nhorido Dzokunyikadzimu*, comp. Ticha Jongwe, 169–170. Gweru: Mambo Press.

Madombi, F. 1983b. "Shiri Kubhunuruka." In *Nehanda Nyakasikana: Nhorido Dzokunyikadzimu*, comp. Ticha Jongwe, 171–172. Gweru: Mambo Press.

Mahachi, M. 1983. "Tauya nezviyo izvi." In *Nehanda Nyakasikana: Nhorido Dzokunyikadzimu*, comp. Ticha Jongwe, 120. Gweru: Mambo Press.

Majaya, E. 1983a. "Regedzai Mwana." In *Nehanda Nyakasikana: Nhorido Dzokunyikadzimu*, comp. Ticha Jongwe, 282–283. Gweru: Mambo Press.

Majaya, E. 1983b. "Kudzinga Zimutsamurimo." In *Nehanda Nyakasikana: Nhorido Dzokunyikadzimu*, comp. Ticha Jongwe, 331–334. Gweru: Mambo Press.

Majoto, L. 1983. "Mudzimu Mukuru Gokorudzi." In *Nehanda Nyakasikana: Nhorido Dzokunyikadzimu*, comp. Ticha Jongwe, 44–46. Gweru: Mambo Press.

Manyimbiri, F. 1983. "Tinoisa Mhuri Mumaoko Enyu." In *Nehanda Nyakasikana: Nhorido Dzokunyikadzimu*, comp. Ticha Jongwe, 133–135. Gweru: Mambo Press.

Mashiringwane, E. 1983a. "Heri Doro Rako Sakurima." In *Nehanda Nyakasikana: Nhorido Dzokunyikadzimu*, comp. Ticha Jongwe, 117–118. Gweru: Mambo Press.

Mashiringwane, E. 1983b. "Chengetai Mwana Wenyu." In *Nehanda Nyakasikana: Nhorido Dzokunyikadzimu*, comp. Ticha Jongwe, 142–143. Gweru: Mambo Press.

Matandaudhle, J. 1983. "Chengetai Mhuri Ino." In *Nehanda Nyakasikana: Nhorido Dzokunyikadzimu*, comp. Ticha Jongwe, 105–107. Gweru: Mambo Press..

Mateta, S. 1983. "Muri Mudzimu Wembeva Here?" In *Nehanda Nyakasikana: Nhorido Dzokunyikadzimu*, comp. Ticha Jongwe, 272–273. Gweru: Mambo Press.

Mugwanda, P. 1983. "Hero Doro Renyu, Moyo." In *Nehanda Nyakasikana: Nhorido Dzokunyikadzimu*, comp. Ticha Jongwe, 98–100. Gweru: Mambo Press.

Mukute, F. F. 1983a. "Madonhwe Oupenyu I." In *Nehanda Nyakasikana: Nhorido Dzokunyikadzimu*, comp. Ticha Jongwe, 64–66. Gweru: Mambo Press.

Mukute, F. F. 1983b."Titarirei." In *Nehanda Nyakasikana: Nhorido Dzokunyikadzimu*, comp. Ticha Jongwe, 108–109. Gweru: Mambo Press.

Nyamukondiwa, W. 1983. "Vamasango Nyawada." In *Nehanda Nyakasikana: Nhorido Dzokunyikadzimu*, comp. Ticha Jongwe, 163–164. Gweru: Mambo Press.

Shumba, T. 1983a. "Chigarai Pamuzukuru Wenyu." In *Nehanda Nyakasikana: Nhorido Dzokunyikadzimu*, comp. Ticha Jongwe, 137–138. Gweru: Mambo Press.

Shumba, T. 1983b. "Tambirai Chibereko." In *Nehanda Nyakasikana: Nhorido Dzokunyikadzimu*, comp. Ticha Jongwe, 147–148. Gweru: Mambo Press.

Shumba, T. 1983c. "Kudzorwa kwaBaba vangu Mumusha." In *Nehanda Nyakasikana: Nhorido Dzokunyikadzimu*, comp. Ticha Jongwe, 152–153. Gweru: Mambo Press.

Shumba, T. 1983d. "Mhuri dzanzwa nenhomba." In *Nehanda Nyakasikana: Nhorido Dzokunyikadzimu*, comp. Ticha Jongwe, 166–167. Gweru: Mambo Press.

Shumba, T. 1983e. "Aguta Haaoneki." In *Nehanda Nyakasikana: Nhorido Dzokunyikadzimu*, comp. Ticha Jongwe, 168–169. Gweru: Mambo Press..

Shumba, T. 1983f. "Kukumikidza Mwana Wechikoro Kuvadzimu." In *Nehanda Nyakasikana: Nhorido Dzokunyikadzimu*, comp. Ticha Jongwe, 182–184. Gweru: Mambo Press.

Shumba, T. 1983g. "Kusatenda Uroyi." In *Nehanda Nyakasikana: Nhorido Dzokunyikadzimu*, comp. Ticha Jongwe, 179–180. Gweru: Mambo Press.

Shumba, T. 1983h. "Rupango Rwotomupa Urwu." In *Nehanda Nyakasikana: Nhorido Dzokunyikadzimu*, comp. Ticha Jongwe, 195–196. Gweru: Mambo Press.

Shumba, T. 1983i. "Kungurira Uroyi." In *Nehanda Nyakasikana: Nhorido Dzokunyikadzimu*, comp. Ticha Jongwe, 351–352. Gweru: Mambo Press.

Vhurinosara, L. 1983a. "Oisaiwo Mumvuri Muzukuru Wenyu." In *Nehanda Nyakasikana: Nhorido Dzokunyikadzimu*, comp. Ticha Jongwe, 178–179. Gweru: Mambo Press.

Vhurinosara, L. 1983b. "Nditaririreiwo Muvatorwa." In *Nehanda Nyakasikana: Nhorido Dzokunyikadzimu*, comp. Ticha Jongwe, 181–182. Gweru: Mambo Press.

Company Records

Transvaal Chamber of Mines: Annual Report for the Year Ended 1897. 1897. Johannesburg: The Chamber.

Transvaal Chamber of Mines: Annual Report for the Year Ended 1912. 1912. Johannesburg: The Chamber.

Transvaal Chamber of Mines: Annual Report for the Year Ended 1914. 1914. Johannesburg: The Chamber.

Articles and Books

"447 kg of Ivory Smuggled to Dubai." 2013. *The Sunday Mail* (September 22). Available at http://www.sundaymail.co.zw.

"African Grand Tour: Fascinating New Prospects for South African Motorists." 1937. *British South Africa Annual* (December): 96.

Anderson, C. J. 1856. *Lake Ngami, or, Explorations and Discoveries*. London: Hurst and Blackett.

"Annual Report of the Chief Entomologist 1921." 1921. *Bulletin of Entomological Research* 2:315.

Baines, Thomas. 1877. *The Gold Regions of South Eastern Africa*. London: Edward Stanford.

Bullock, C. 1950. *The Mashona and the Matabele*. Cape Town: Juta.

Bulpin, T. V. 1954. *The Ivory Trail*. Cape Town: T. V. Bulpin.

Buntting, E. 1949. "The Genet, Enemy of Snakes." *African Wild Life* 2 (3): 29.

Capstick, Peter. 1988. *The Last Ivory Hunter*. New York: St. Martin's Press.

Cary, Eric G. 1954. "Extracts from Boys Diaries." *Rhodesian Schools Exploration Society Report: Lundi Expedition*. Salisbury: National Museums and Monuments.

Chara, Tendai. 2013. "Elephant Killings: A War within a War." *The Sunday Mail* (October 6). Available at http://www.sundaymail.co.zw.

Chinodya, Shimmer. 1989. *Harvest of Thorns*. Harare: Baobab.

"The Civilizing Influence of Roads: Intercommunication Developments an Urge to Advancement." 1929–1930. *British South Africa Annual*: 144.

"Cops Attacked by Villagers in Tsholotsho." 2013. *NewsdzeZimbabwe* (March 31). Available at http://www.newsdzezimbabwe.co.uk.

Cox, Peter. 2013. "Elephants Killed by Cyanide Reveal Alarming Innovation in Poaching Tactics." *Voice of America* (October 1). Available at http://www.voanews .com.

"Cyanide Poachers Remanded in Custody." 2013. *Daily News* (October 1). Available at http://www.dailynews.co.zw.

das Neves, Fernandes. 1879. *Fernandes. A Hunting Expedition to the Transvaal*. London: G. Bell and Sons.

"Development of Game Reserves in the Colony: Official Statement to Publicity Conference: 'Tsetse Fly Obstacle to Extension of the Kruger National Park.'" 1937. *The Rhodesia Herald* (August 13).

Editor. 1926. "Native Iron Workers." *NADA* 4:53.

Erskine, St. Vincent. 1875. "A Journey to Umzila, in South Eastern Africa." *Journal of the Royal Geographical Society* 19:110–134.

Fairchild, Hoxie Neale. 1928. *The Noble Savage: A Study in Romantic Naturalism.* New York: Columbia University Press.

Gielgud, Val. 1908. "Notes on Game in Southern Rhodesia." *Journal of the Society for the Preservation of the Wild Fauna of the Empire* 4:35.

Gillmore, Parker. 1890. *Through Gasa Land, and the Scene of the Portuguese Aggression: The Journey of a Hunter in Search of Gold and Ivory.* London: Harrison and Sons.

Hall, Ivan C., and Richard W. Whitehead. 1927. "A Pharmaco-Bacteriologic Study of African Poisoned Arrows." *Journal of Infectious Diseases* 41 (1): 51–69.

Hatton, J. S. 1967. "Notes on Makalanga Iron Smelting." *NADA* 9 (4): 39–42.

"Health Services at Risk." 1977. *Focus* (September).

"The Hellish Life." 1978. *The Voice* (November 25).

Hutcheon, Dr. 1896. "The 'Rinderpest': What It Is with Symptoms and Causes: An Interview with Dr. Hutcheon, M.R.C.V.S." *African Review* (May 23): 1027.

"Hwange Clean-up, De-Tox Underway." 2013. *The Herald* (September 30). Available at http://www.herald.co.zw.

"Increasing Market for Motor Vehicles: New Cars Purchased in September." 1934. *The African World* (December 8): 294.

Jackson, A. P. 1950. "Native Hunting Customs." *NADA* 27:39–41.

Junod, Henri A. 1927. *The Life of a South African Tribe.* 2 vols. Neuchatel: Attinger.

"Kangai Now a Leading Distributor of Chinese Medicines in Zimbabwe." *NewsdzeZimbabwe.* (May 16, 2012) Available at www.newsdzezimbabwe.co.uk.

Kinnell, Sheila B. 1958. "Stalking the Hunters by Car." *African Wild Life* 12 (3) (September): 155–157.

Kirk, John. 1865. "On the 'Tsetse' Fly of Tropical Africa." *Journal of the Linnean Society* 8:149–156.

Kirk, John. 1907. "The Tsetse Fly as a Disease-Carrier." *Journal of the Society of the Preservation of the Wild Fauna of the Empire* 3:45–46.

Kruger National Park. 1933. "Fighting Sleeping Sickness." *African Wild Life* (July 29): 64.

Lloyd, Elaine M. 1925. "Mbava." *NADA* 3:62–63.

Low, Chris. 2007. "Khoisan Wind: Hunting and Healing." *Journal of the Royal Anthropological Institute* 13:S71–S90.

MacVicar, Neil. 1917. "Kafir Poisoning." *South African Medical Record* 15 (1): 3.

Magadza, Chris. 2013. "New Techniques of Tackling Poaching." *The Sunday Mail* (October 6). Available at http://www.sundaymail.co.zw.

Matibe, Phil. 2009. *Madhinga Bucket Boy: From Boyhood in Colonial Rhodesia to Manhood in Zimbabwe*. Dallas: Mbedzi Publishing.

Mhlanga, Wilson. 1948. "The Story of Ngwaqazi, 2: The History of the Amatshangana." *NADA* 25:70–73.

Millais, John G. 1895. *A Breath from the Veldt*. London: Henry Sotheran.

Montagu, John Scott. 1896. "The Matabeli Rising and Its Causes." *African Review* (June 6): 1121.

"More Motor-Cars in Southern Rhodesia: Preponderance of American Models." 1936. *The African World* (April 18): 21.

"Motor Touring in Rhodesia." 1931. *British South Africa Annual* (December): 139.

Mudzungairi, Wisdom. 2013. "Is Kasukuwere Full of All Sound and Fury?" *Newsday* (September 16). Available at https://www.newsday.co.zw.

Mukarati, Levi. 2013."What's in a Tusk?" *The Sunday Mail* (September 29). Available at http://www.sundaymail.co.zw.

Muponde, Richard. 2013. "Police Officers in Court over Ivory Bribes." *The Standard* (October 6). Available at http://www.thestandard.co.zw.

"New Motor Vehicles Registered." 1935. *The African World* (March 23): 355.

"The New Zambesi Bridge Completed." 1934. *The African World* (November 3): 103.

Nkomo, Joshua. 1984. *Nkomo: The Story of My Life*. London: Methuen.

Odendaal, P. J. 1941. "The Bow and Arrow in Southern Rhodesia." *NADA* 18:23–24.

Parkhurst, D. C. H. 1948. "Beliefs about the Elephant." *NADA* 25:49.

Partridge, G. A. 1958. "Traps a Menace to Wild Life." *African Wild Life* 12 (3) (September): 117.

Philpot, J. 1954. "Anthropology." *Rhodesian Schools Exploration Society Report: Lundi Expedition*. Salisbury: National Museums and Monuments.

Posselt, F. W. T. 1935. *Fact and Fiction; A Short Account of the Natives of Southern Rhodesia*. Bulawayo: Government of Southern Rhodesia.

Preller, Gustav. 1922. *Voortrekker Mense III*. Kaapstad and Bloemfontein: de Naionale Pers.

"Publicity Collaboration in Southern Africa Urged at National Conference in Salisbury—'Obstacles to Rhodesian Extension of Kruger Park.'" 1937. *The Rhodesia Herald* (August 13).

Rahm, U. 1960. "The Pangolins of West and Central Africa." *African Wild Life* 14 (4): 271–275.

"Rearrangement of Portfolios." 1933. *The African World* (October 14): 432.

"The Results of Rinderpest." 1897. *The African Review* (September 4): 459.

"Rhodesian Tortures Are Exposed: Police Arrest Priests." 1977. *The Observer* (September 4).

"The Rinderpest." 1896. *African Review* (November 28): 406.

Schapera, I. 1927. "Bows and Arrows of the Bushmen." *Man* 27:113–117.

Scott, Peter. 1954. "Migrant Labor in Southern Rhodesia." *Geographical Review* 44 (1): 29–48.

Selous, Frederick C. 1881. *A Hunter's Wanderings in Africa.* London: Richard Bentley & Son.

"Shooting Game from a Motor-Lorry: Foreign Princes Fined in Kenya." 1934. *The African World* (March 17): 344.

Simpson, D., and S. Booysens. 1958. "Animal and Plant Ecology Report." *Rhodesian Schools Exploration Society Report: Mateke Expedition.* Salisbury: National Museums and Monuments.

Snowden, A. E. 1940. "Some Technological Notes on Weapons and Implements Used in Mashonaland." *NADA* 17:62–70.

Stayt, Hugh A. 1931. *The Bavenda.* London: Oxford University Press.

"Stock Restrictions Withdrawn." 1933. *The African World* (April 22): 22.

Strover, P. 1958. "In and Around Camp—General." *Rhodesian Schools Exploration Society Report: Mateke Expedition.* Salisbury: National Museums and Monuments.

Struben, Hendrik. 1920. *Recollections of Adventures, 1850–1911.* Cape Town: T. M. Miller.

Swynnerton, Charles F. M. 1921. "An Examination of the Tsetse Problem in North Mossurise, Portuguese East Africa." *Bulletin of Entomological Research* 11:315–385.

Thompson, Louis C. 1949. "Ingots of Native Manufacture." *NADA* 26:7–19.

Thomson, Ron. 2001. *The Adventures of Shadrek: Southern Africa's Most Infamous Elephant Poacher.* Long Beach, CA: Safari Press.

Tomlinson, A. J. 1936. "A Motoring Holiday in Rhodesia." *British South Africa Annual* (December): 147–149.

"Traditional Medicine." 2003. WHO Fact Sheet No.134. Available at http://www.who.int/mediacentre/factsheets.

UNCTAD. 1972. *Guidelines for the Study of the Transfer of Technology to Developing Countries*. Geneva: UNCTAD Secretariat.

Wallis, J. P. R., ed. 1945. *The Matabele Journals of Robert Moffat, 1829–1860*. 2 vols. London: Chatto & Windus.

Wallis, J. P. R. 1954. *The Southern African Diaries of Thomas Leask, 1865–1870*. London: Chatto and Windus.

"We Poisoned Jumbos for a Living, Villagers Confess." 2013. *The Herald* (October 5). Available at http://www.herald.co.zw.

Wright, Allan. 1972. *Valley of the Ironwoods*. Cape Town: T. V. Bulpin.

Wright, Allan. 1976. *Grey Ghosts at Buffalo Bend*. Cape Town: T. V. Bulpin.

"Zimbabwe Poachers Kill 80 Elephants, Poisoning Water Holes with Cyanide." 2013. *The Guardian* (September 25). Available at http://www.theguardian.com.

Zumpt, F. 1960. "A Trip to a Game Paradise 3. The Warthog and Its Parasites." *African Wild Life* 14 (1): 29–33.

Secondary Sources

Abraham, Itty. 1998. *The Making of the Indian Atomic Bomb: Science, Secrecy and the Postcolonial State*. London: Zed Books.

Achebe, Chinua. 1978. "An Image of Africa." *Research in African Literatures* 9 (1): 1–15.

Adas, Michael. 1989. *Machines as the Measure of Men: Science, Technology, and Ideologies of Western Dominance*. Ithaca and London: Cornell University Press.

Adas, Michael. 2009. *Dominance by Design: Technological Imperatives and America's Civilizing Mission*. Cambridge, MA: Harvard University Press.

Adey, Peter. 2009. *Mobility*. Abingdon: Routledge.

Agamben, Giorgio. 1998. *Homo Sacer: Sovereign Power and Bare Life*, trans. Daniel Heller-Roazen. Palo Alto: Stanford University Press.

Agamben, Giorgio. 2005. *State of Exception*, trans. Kevin Attell. Chicago: University of Chicago Press.

Akrich, M. 1992. "The De-Scription of Technical Objects." In *Shaping Technology, Building Society: Studies in Sociotechnical Change*, ed. Wiebe Bijker and John Law, 205–224. Cambridge, MA: MIT Press.

Alexander, Jocelyn, JoAnn McGregor, and Terence Ranger. 2000. *Violence & Memory: One Hundred Years in the "Dark Forests" of Matabeleland.* Oxford: James Currey.

Anderson, David M. 1984. "Depression, Dust Bowl, Demography and Drought: The Colonial State and Soil Conservation in East Africa during the 1930s." *African Affairs* 83 (332): 321–343.

Anderson, David, and Richard Grove, eds. 1987. *Conservation in Africa: People, Policies and Practice.* Cambridge: Cambridge University Press.

Anderson, Warwick. 1992. "'Where Every Prospect Pleases and Only Man Is Vile': Laboratory Medicine as Colonial Discourse." *Critical Inquiry* 18:506–529.

Appadurai, Arjun. 1996. *Modernity at Large: Cultural Dimensions of Globalization.* Minneapolis: University of Minnesota Press.

Atmore, Antony, Mutero Chirenje, and Stan Mudenge. 1971. "Firearms in South Central Africa." *Journal of African History* 12 (4): 545–556.

Balint, P. J., and J. Mashinya. 2008. "CAMPFIRE during Zimbabwe's National Crisis: Local Impacts and Broader Implications for Community-based Wildlife Management." *Society & Natural Resources* 21 (9): 783–796.

Banister, David. 2005. "Reducing Travel by Design: What about Change over Time?" In *Spatial Planning, Urban Form and Sustainable Transport*, ed. Katie Williams, 102–119. Aldershot, UK: Ashgate.

Bannerman, J. H. 1981. "Hlengweni: The History of the Hlengwe of the Lower Save and Lundi Rivers, from the Late Eighteenth to the Mid-Twentieth Century." *Zimbabwean History* 12:1–45.

Bauman, Zygmunt. 2000. *Liquid Modernity.* Cambridge: Polity.

Bayart, Jean-Francois. 2000. "Africa in the World: A History of Extraversion." *African Affairs* 99:217–267.

Beach, David. 1974. "Ndebele Raiders and Shona Power." *Journal of African History* 15 (4): 633–651.

Beinart, William. 1984. "Soil Erosion, Conservationism and Ideas about Development: A Southern African Exploration, 1900–1960." *Journal of Southern African Studies* 11 (1): 52–83.

Beinart, William. 2000. "African History and Environmental History." *African Affairs* 99 (395): 269–302.

Bhabha, Homi. 1996. "Unsatisfied: Notes on Vernacular Cosmopolitanism." In *Text and Nation*, ed. Laura Garcia-Morena and Peter C. Pfeifer, 191–207. London: Camden House.

Bhebe, Ngwabi. 1979. *Christianity and Traditional Religion in Western Zimbabwe*. London: Longman.

Bhila, Hoyini. 1982. *Trade and Politics in a Shona Kingdom*. London: Longman.

Bille, Mikkel, Frida Hastrup, and Tim Flohr Sørensen, eds. 2010. *An Anthropology of Absence: Materializations of Transcendence and Loss*. New York: Springer.

Bloor, David. 1976. *Knowledge and Social Imagery*. Chicago: University of Chicago Press.

Bonner, Philip. 1983. *Kings, Commoners and Concessionaires: The Evolution and Dissolution of the Nineteenth-Century Swazi State*. Cambridge: Cambridge University Press.

Bowker, Geoffrey, and Susan Star. 1999. *Sorting Things Out: Classification and Its Consequences*. Cambridge, MA: MIT Press.

Brown, D. M. 1983. "The Noble Savage in Anglo-Saxon Colonial Ideology, 1950–1980: 'Masai' and 'Bushmen' in Popular Fiction." *English in Africa* 10 (2): 55–77.

Brown, Karen. 2007. "Poisonous Plants, Pastoral Knowledge and Perceptions of Environmental Change in South Africa, c. 1880–1940." *Environmental History* 13 (3): 307–332.

Brown, Karen, and Daniel Gilfoyle eds. 2007. "Livestock Diseases and Veterinary Science." *South African Historical Journal* 58:1–141.

Callon, Michel. 1986. "Some Elements of a Sociology of Translation: Domestication of the Scallops and the Fishermen of St. Brieuc Bay." In *Power, Action and Belief: A New Sociology of Knowledge*, ed. John Law, 196–233. London: Routledge & Kegan Paul.

Campbell, Gwyn. 1987. "The Adoption of Autarky in Imperial Madagascar, 1820–1835." *Journal of African History* 28:395–409.

Carruthers, Jane. 1995. *The Kruger National Park: A Social and Political History*. Pietermaritzburg: University Natal Press.

Castells, Manuel. 1996. *The Rise of the Network Society*. Oxford: Wiley-Blackwell.

Césaire, Aime. 1955/2000. *Discourse on Colonialism*, trans. Joan Pinkham. New York: Monthly Review Press.

Chakrabarty, Dipesh. 2000. *Provincializing Europe: Postcolonial Thought and Historical Difference*. Princeton: Princeton University Press.

Chauke, Matsilele. 1985. "Ndambakuwa's Administration, Chibi District, 1896–1914." BA honors dissertation, University of Zimbabwe.

Chikowero, Mhoze. 2007. "Subalternating Currents: Electrification and Power Politics in Colonial Bulawayo, Colonial Zimbabwe, 1894–1939." *Journal of Southern African Studies* 33 (2): 287–306.

Chilundo, Arlindo. 1995. "The Economic and Social Impact of the Rail and Road Transportation Systems in the Colonial District of Moçambique (1900–1961)." PhD dissertation, University of Minnesota.

Chirenje, J. Mutero. 1973. "Portuguese Priests and Soldiers in Zimbabwe, 1560–1572: The Interplay between Evangelism and Trade." *International Journal of African Historical Studies* 6 (1): 36–48.

Chiteji, Fran. 1979. "The Development and Socio-Economic Impact of Transportation in Tanzania, 1884–Present." PhD dissertation, Michigan State University.

Cobbing, Julian. 1973. "Lobengula, Jameson and the Occupation of Mashonaland, 1890." *Rhodesian History* 4:40–43.

Cohen, David W. 1977. *Womunafu's Bunafu*. Princeton: Princeton University Press.

Conrad, Joseph. 1902. *Heart of Darkness*. New York: Harper and Brothers.

Cronon, William. 1983. *Changes in the Land: Indians, Colonists, and the Ecology of New England*. New York: Hill and Wang.

Crosby, Alfred. 1993. *Ecological Imperialism: The Biological Expansion of Europe, 900–1900*. New York: Cambridge University Press.

Crush, Jonathan, and Daniel Tevera, eds. 2010. *Zimbabwe's Exodus: Crisis, Migration, Survival*. Cape Town: Southern African Migration Programme.

Curtin, P. D. 1975. *Economic Change in Pre-colonial Africa*. Madison: University of Wisconsin Press.

Cwerner, Saulo, Sven Kesselring, and John Urry, eds. 2009. *Aeromobilities*. London and New York: Routledge.

Daneel, Martinus. 1995. *Guerrilla Snuff*. Harare: Baobab.

Darwin, Charles. 1859. *On the Origin of Species*. London: John Murray.

Delius, Peter, and Stanley Trapido. 1983. "*Inboekselings* and *Oorlams*: The Creation and Transformation of a Servile Class." In *Town and Countryside in the Transvaal*, ed. Belinda Bozzoli, 53–88. Johannesburg: Raven.

Dinerstein, Joel. 2006. "Technology and Its Discontents: On the Verge of the Posthuman." *American Quarterly* 58 (3): 569–595.

Diouf, Mamadou. 2000. "The Senegalese Murid Trade Diaspora and the Making of a Vernacular Cosmopolitanism." *Public Culture* 12 (3): 679–702.

Divall, Collin. 2010. "Mobilizing the History of Technology." *Technology and Culture* 51 (4): 938–960.

Divall, Colin, and George Revill. 2005. "Cultures of Transport: Representation, Practice and Technology." *Journal of Transport History* 26 (1): 99–117.

Divall, Colin, and George Revill. 2006. "No Turn Needed: A Reply to Michael Freeman." *Journal of Transport History* 27 (1): 144–149.

Drinkwater, Mike. 1989. "Technical Development and Peasant Impoverishment: Land Use Policy in Zimbabwe's Midlands Province." *Journal of Southern African Studies* 15 (2): 287–305.

Driver, Thackwray. 1999. "Anti-Erosion Policies in the Mountain Areas of Lesotho: The South African Connection." *Environmental History* 5 (1): 1–25.

Dumett, Raymond E. 1983. "African Merchants of the Gold Coast, 1860–1905—Dynamics of Indigenous Entrepreneurship." *Comparative Studies in Society and History* 25 (4): 661–693.

Echenberg, Myron J. 1971. "Late Nineteenth-Century Military Technology in Upper Volta." *Journal of African History* 12 (2): 241–254.

Edgerton, David. 2007. *The Shock of the Old: Technology and Global History Since 1900.* Oxford: Oxford University Press.

Eisenstadt, S. N. 2000. "Multiple Modernities." *Daedalus* 129 (1): 1–29.

Ellert, Henrik. 1984. *The Material Culture of Zimbabwe.* Harare: Longman.

Fabian, Johannes. 1983. *Time and the Other: How Anthropology Makes Its Object.* New York: Columbia University Press.

Fairhead, James, and Melissa Leach. 1996. *Misreading the African Landscape: Society and Ecology in a Forest-Savanna Mosaic.* Cambridge: Cambridge University Press.

Fanon, Frantz. 1952/1967. *Black Skin, White Masks,* trans. Charles L. Markmann. New York: Grove Press.

Feierman, Steve, and John Janzen, eds. 1992. *The Social Basis for Healing and Healing in Africa.* Berkeley: University of California Press.

Ferguson, James. 2006. *Global Shadows: Africa in the Neoliberal World Order.* Durham, NC: Duke University Press.

Fisher, Humphrey, and Virginia Rowland. 1971. "Firearms in the Central Sudan." *Journal of African History* 12 (2): 215–239.

Fleck, Ludwik. 1935/1979. *The Genesis and Development of a Scientific Fact.* Ed. T. J. Trenn and R. K. Merton. Chicago: University of Chicago Press.

Ford, John. 1971. *The Role of the Trypanosomiases in African Ecology: A Study of the Tsetse Fly Problem*. Oxford: Clarendon Press.

Freeman, Michael. 2006. "'Turn If You Want To': A Comment on the 'Cultural Turn' in Divall and Revill's 'Cultures of Transport.'" *Journal of Transport History* 27 (1): 138–143.

Fyle, C. Magbaily. 1987. "Culture, Technology and Policy in the Informal Sector: Attention to Endogenous Development." *Africa* 57 (4): 498–509.

Gaonkar, Dilip. 2001. *Alternative Modernities*. Durham, NC: Duke University Press.

Geschiere, Peter. 1997. *The Modernity of Witchcraft: Politics and the Occult in Postcolonial Africa*. Charlottesville, London: University Press of Virginia.

Gewald, Jan-Bart, Sabine Luning, and K. van Walraven, eds. 2009. *The Speed of Change: Motor Vehicles and People in Africa, 1890–2000*. Leiden: Brill.

Giddens, Anthony. 2002. *Where Now for New Labour?* Cambridge: Polity.

Gilfoyle, Dan. 2003. "Veterinary Research and the African Rinderpest Epizootic: The Cape Colony, 1896–1898." *Journal of Southern African Studies* 29 (1): 133–154.

Goody, Jack. 1971. *Technology, Tradition, and the State in Africa*. Oxford: Oxford University Press.

Green, Maia, and David Hulme. 2005. "From Correlates and Characteristics to Causes: Thinking about Poverty from a Chronic Poverty Perspective." *World Development* 33 (6): 867–879.

Guelke, Leonard, and Robert Shell. 1992. "Landscapes of Conquest: Frontier Water Alienation and Khoikhoi Strategies of Survival." *Journal of Southern African Studies* 18 (4): 803–824.

Guyer, Jane. 2004. *Marginal Gains: Monetary Transactions in Atlantic Africa*. Chicago: University of Chicago Press.

Hardt, Michael, and Antonio Negri. 2001. *Empire*. Cambridge, MA: Harvard University Press.

Harries, Patrick. 1981. "Slavery, Social Incorporation and Surplus Extraction: The Nature of Free and Unfree Labour in South-east Africa." *Journal of African History* 22 (3): 309–330.

Hartman, Saidiya. 2007. *Lose Your Mother: A Journey along the Atlantic Slave Route*. New York: Farrar, Straus and Giroux.

Headrick, Daniel R. 1981. *The Tools of Empire: Technology and European Imperialism in the Nineteenth Century*. New York: Oxford University Press.

Headrick, Daniel R. 2010. *Power Over Peoples: Technology, Environments, and Western Imperialism, 1400 to the Present*. Princeton: Princeton University Press.

Hecht, Gabrielle. 2012. *Being Nuclear: Africans and the Global Uranium Trade*. Cambridge, MA: MIT Press.

Hegel, Georg F. 1837/2007. *The Philosophy of History*. New York: Cosimo, Inc.

Held, David, Anthony McGrew, David Goldblatt, and Jonathan Perraton. 1999. *Global Transformation: Politics, Economics, Culture*. Stanford: Stanford University Press.

Holston, James, ed. 1999. *Cities and Citizenship*. Durham, NC: Duke University Press.

Hopkins, A. G. 1973. *An Economic History of West Africa*. London: Longmans.

Jewsiewicki, Bogumil. 1991. "The Archaeology of Invention: Mudimbe and Postcolonialism." *Callaloo* 14 (4): 961–968.

Jirón, Paola. 2010. "Mobile Borders in Urban Daily Mobility Practices in Santiago de Chile." *International Political Sociology* 4 (1): 66–79.

Jules-Rosette, Benetta. 1991. "*Speaking about Hidden Times*: The Anthropology of V. Y. Mudimbe." *Callaloo* 14 (4): 944–960.

Kanbur, Ravi, and Lyn Squire. 2001. "The Evolution of Thinking about Poverty: Exploring the Interactions." In *Frontiers of Development Economics*, ed. G. Meier and J. Stiglitz, 183–226. New York: Oxford University Press.

Kea, R. A. 1971. "Firearms and Warfare on the Gold and Slave Coasts from the Sixteenth to the Nineteenth Centuries." *Journal of African History* 12 (2): 185–213.

Kjekshus, Helge. 1996. *Ecology Control and Economic Development in East African History*. London: James Currey.

Kosmin, B. A. 1975. "'Freedom, Justice and Commerce': Some Factors Affecting Asian Trading Patterns in Southern Rhodesia, 1897–1942." *Rhodesian History* 6:15–32.

Lan, David. 1985. *Guns and Rain: Guerrillas and Spirit Mediums in Zimbabwe*. London: James Currey.

Latour, Bruno. 1987. *Science in Action: How to Follow Scientists and Engineers through Society*. Cambridge, MA: Harvard University Press.

Latour, Bruno, and Steve Woolgar. 1979. *Laboratory Life: The Social Construction of Scientific Facts*. Princeton: Princeton University Press.

Legassick, Martin. 1966. "Firearms, Horses and Samorian Army Organization 1870–1898." *Journal of African History* 7 (I): 95–115.

Lenin, Vladimir. 1917. *Imperialism, the Highest Stage of Capitalism*. Petrograd: Zhzn i, Znaniye Publishers.

Levi-Strauss, Claude. 1966. *The Savage Mind*. Chicago: University of Chicago Press.

Liesegang, Gerhard. 1967. "Beitràge zur Geschichte des Reiches der Gaza Nguni im súdlicken Moçambique, 1820–1895." PhD dissertation, University of Cologne.

Liesegang, Gerhard. 1975. "Aspects of Gaza Nguni History, 1821–1897." *Rhodesian History* 6:1–14.

Lindsey, P. A., et al. 2011. "Dynamics and Underlying Causes of Illegal Bushmeat Trade in Zimbabwe." *Oryx* 45 (1): 84–95.

Lipschutz, Mark R., and R. Kent Rasmussen. 1989. *Dictionary of African Historical Biography*. Berkeley: University of California Press.

MacKenzie, John. 1988. *The Empire of Nature: Hunting, Conservation and British Imperialism*. Manchester: Manchester University Press.

MacNaughten, Phil, and John Urry. 2000. "Bodies in the Woods." *Body & Society* 6:166–182.

Malowist, Marian. 1966. "Le commerce d'or et d'eclaves au Soudan Occidental." *Africana Bulletin* 4:49–93.

Marks, Shula, and Anthony Atmore, eds. 1980. *Economy and Society in Pre-Industrial South Africa*. London: Longman.

Marks, Shula, and Richard Rathbone, eds. 1982. *Industrialisation and Social Change in South Africa: African Class Formation, Culture and Consciousness, 1870–1930*. London: Longman.

Marks, Stuart A. 1976. *Large Mammals and a Brave People: Subsistence Hunters in Zambia*. Seattle: University of Washington Press.

Marx, Karl. 1867/1954. *Capital*, vol. I. Moscow: Foreign Languages Publishing House.

Marx, Leo. 2010. "Technology: The Emergence of a Hazardous Concept." *Technology and Culture* 51 (3): 561–577.

Masolo, D. A. 1991. "An Archaeology of African Knowledge." *Callaloo* 14 (4): 998–1011.

Matose, Frank. 2006. "Co-management Options for Reserved Forests in Zimbabwe and Beyond: Policy Implications of Forest Management Strategies." *Forest Policy and Economics* 8 (4): 363–374.

Mavhunga, Clapperton Chakanetsa. 2001. "Sold Down the River? Forced Resettlement and Landscape Transformation: Lessons from the Kariba Dam, 1950–63." Unpublished paper, University of Zimbabwe History Department.

Mavhunga, Clapperton Chakanetsa. (2003). "Firearms Diffusion, Exotic and Indigenous Knowledge Systems in the Lowveld Frontier, South Eastern Zimbabwe, 1870–1920." *Comparative Technology Transfer and Society* 1 (August): 201–231.

Mavhunga, Clapperton C. 2007a. "Navigating Boundaries of Urban/Rural Migration in Southeastern Zimbabwe, 1890s–1920s." In *African Agency and European Colonialism: Latitudes of Negotiation and Containment*, ed. F. J. Kolapo and K. O. Akurang-Parry, 121–140. Lanham, MD: University Press of America.

Mavhunga, Clapperton Chakanetsa. 2007b. "Even the Rider and a Horse are a Partnership: a Reply to Vermeulen and Sheil." *Oryx* 41 (4): 441–442.

Mavhunga, Clapperton Chakanetsa, and Wolfram Dressler. 2007. "On the Local Community: The Language of Disengagement?" *Conservation & Society* 5 (1): 44–59.

Mavhunga, Clapperton Chakanetsa. 2011a. "On Vermin Beings: On Pestiferous Animals and Human Game." *Social Text* 106:151–176.

Mavhunga, Clapperton Chakanetsa. 2011b. "A Plundering Tiger with its Deadly Cubs? The USSR and China in the Engineering of a 'Zimbabwean Nation,' 1945–2009." In *Entangled Geographies: Empire and Technopolitics in the Global Cold War*, ed. G. Hecht, 231–266. Cambridge, MA: MIT Press.

Mavhunga, Clapperton Chakanetsa. 2012. "Which Mobility for (Which) Africa? Beyond Banal Mobilities." *T²M Yearbook*: 73–84.

Mavhunga, Clapperton Chakanetsa. 2013. "'Cidades Esfumacadas': Energy and the Rural-Urban Connection in Mozambique." *Public Culture* 25 (2): 261–271.

Mavhunga, Clapperton Chakanetsa, and Marja Spierenburg. 2007. "A Finger on the Pulse of the Fly: Hidden Voices of Colonial Anti-Tsetse Science on the Rhodesia and Mozambique Borderlands, 1945–1956." *South African Historical Journal* 58:117–141.

Mavhunga, Clapperton Chakanetsa, and Marja Spierenburg. 2009. "Transfrontier Talk, Cordon Politics: The Early History of the Great Limpopo Transfrontier Park in Southern Africa, 1925–1940." *Journal of Southern African Studies* 35:715–735.

Mazarire, Gerald. 2005. "Defence Consciousness as Way of Life: 'The Refuge Period' and Karanga Defence Strategies in the 19th Century." *Zimbabwean Prehistory* 25:19–26.

Mbembe, Achille. 2002. "African Modes of Self-Writing." *Public Culture* 14 (1): 239–273.

Miller, Duncan, Nirdev Desai, and Julia Lee-Thorp. 2000. "Indigenous Mining in Southern Africa: A Review." *Goodwin Series* 8:91–98.

Mom, Gijs. 2004. "What Kind of Transport History Did We Get? Half a Century of JTH and the Future of the Field." *Journal of Transport History* 24:121–138.

Monson, Jamie. 2009. *Africa's Freedom Railway: How a Chinese Development Project Changed Lives and Livelihoods in Tanzania*. Bloomington, IN: Indiana University Press.

Moore, Henrietta, and Meghan Vaughan. 1994. *Cutting Down Trees: Gender, Nutrition and Agricultural Change in the Northern Province of Zambia, 1890–1990*. London: James Currey.

Mudenge, Stan. 1988. *A Political History of Munhumutapa, c. 1400–1902*. Harare: Zimbabwe Publishing House.

Mudimbe, V. Y. 1973. *L'Autre face du royaume: Une introduction à la critique des langages en folie*. Lausanne: L'Age d'homme.

Mudimbe, V. Y. 1988. *The Invention of Africa*. Bloomington, IN: Indiana University Press.

Mudimbe, V. Y. 1994. *The Idea of Africa*. Bloomington, IN: Indiana University Press.

Murphree, Marshall. 1990. "Communities as Resource Management Institutions." *Gatekeeper Series* SA36:1–14.

Mutwira, Roben. 1989. "Southern Rhodesian Wildlife Policy. 1890–1953: A Question of Condoning Game Slaughter?" *Journal of Southern African Studies* 15 (2): 250–262.

Newell, Stephanie. 2001. "'Paracolonial' Networks: Some Speculations on Local Readership in Colonial West Africa." *Interventions* 3 (3): 336–354.

Ngwerume, E. T., and Cyprian Muchemwa. 2011. "Community Based Natural Resource Management (CBNRM): A Vehicle towards Sustainable Rural Development. The Case of CAMPFIRE in Zimbabwe's Mashonaland West Hurungwe District." *Journal of Emerging Trends in Economics and Management Sciences* 2 (2): 75–82.

Njamnjoh, Francis. 2004. *Rights and the Politics of Recognition in Africa*. London, New York: Zed Books.

Noble, David. 1997. *The Religion of Technology*. New York: Knopf.

Nyevera, R. "Nhanganyaya." 1983. In *Nehanda Nyakasikana: Nhorido Dzokunyikadzimu*, comp. Ticha Jongwe, 15–37. Gweru: Mambo Press, 1983.

Nzaro, Robert, Kosmas Njanike, and Emma Munenerwa. 2011. "An Evaluation of Financial Strategies Used by Companies in the Retail Sector during Recession Period (2000–2010)." *Journal of Business Management and Economics* 2 (1): 22–27.

Odera-Oruka, H. 1983. "Sagacity in African Philosophy." *International Philosophical Quarterly* 23 (4): 383–393.

Ofcansky, Thomas P. 2002. *Paradise Lost: A History of Game Preservation in East Africa*. Morgantown: West Virginia University Press.

Oudshoorn, Nelly, and Trevor Pinch, eds. 2003. *How Users Matter: The Co-Construction of Users and Technology.* Cambridge, MA: MIT Press.

Palmer, Robin. 1977. *Land & Racial Domination in Rhodesia.* Berkeley: University of California Press.

Palmer, Robin. 1990. "Land Reform in Zimbabwe, 1980–1990." *African Affairs* 89:163–181.

Parsons, Neil. 1993. *A New History of Southern Africa.* London: Macmillan.

Peters, Pauline E. 1994. *Dividing the Commons: Politics, Policy, and Culture in Botswana.* Charlottesville: University Press of Virginia.

Philips, John Edward. 2006. "What's New about African History?" *History News Network* (April 6).

Paolini, Albert. 1997. "Globalization." In *At the Edge of International Relations,* ed. Philip Darby, 33–60. London, New York: Continuum.

Phimister, Ian R. 1974. "Alluvial Gold Mining and Trade in Nineteenth-Century South Central Africa." *Journal of African History* 15 (3): 217–228.

Phoofolo, Pule. 1993. "Epidemics and Revolutions: The Rinderpest Epidemic in Late Nineteenth-century Southern Africa." *Past & Present* 138:112–143.

Phoofolo, Pule. 2003. "Face to Face with Famine: The BaSotho and the Rinderpest, 1897–1899." *Journal of Southern African Studies* 29 (2): 503–527.

Pinch, Trevor J., and Wiebe E. Bijker. 1984. "The Social Construction of Facts and Artefacts." *Social Studies of Science* 14:399–441.

Pirie, Gordon. 1982. "The Decivilising Rails: Railways and Underdevelopment in Southern Africa." *Tijdschrift voor Economische en Sociale Geografie* 73 (4): 221–228.

Pirie, Gordon. 1986. "The Cape Colony's 'Railway Protector of Natives,' 1904." *Journal of Transport History* 7:80–92.

Pirie, Gordon. 1987. "African Township Railways and the South African State, 1902–1963." *Journal of Historical Geography* 13 (3): 283–295.

Pirie, Gordon. 1990. "Aviation, Apartheid and Sanctions: Air Transport to and from South Africa, 1945–1989." *GeoJournal* 22 (3): 231–240.

Pirie, Gordon. 1992. "Southern African Air Transport after Apartheid." *Journal of Modern African Studies* 30:341–348.

Pirie, Gordon. 1993a. "Railway Ganging in Southern Africa, c. 1900–37." *Journal of Transport History* 14 (1): 64–76.

Pirie, Gordon. 1993b. "Slaughter by Steam: Railway Subjugation of Ox-wagon Transport in the Eastern Cape and Transkei, 1886–1910." *International Journal of African Historical Studies* 26 (2): 319–343.

Pirie, Gordon. 1993c. "Railways and Labour Migration to the Rand Mines: Constraints and Significance." *Journal of Southern African Studies* 19 (4): 713–730.

Pirie, Gordon. 1997. "Brutish *Bombelas*: Trains for Migrant Gold Miners in South Africa, c. 1900–1925." *Journal of Transport History* 18:31–44.

Pirie, Gordon. 2003. "Cinema and British Imperial Civil Aviation, 1919–1939." *Historical Journal of Film, Radio and Television* 23 (2): 117–131.

Pirie, Gordon. 2004. "Passenger Traffic in the 1930s on British Imperial Air Routes: Refinement and Revision." *Journal of Transport History* 25:63–83.

Pirie, Gordon. 2006. "'Africanisation' of South Africa's International Air Links, 1994–2003." *Journal of Transport Geography* 14:3–14.

Pirie, Gordon. 2008. "Airport Agency: Globalization and (Peri)urbanism." *Alizes* 29:25–36.

Pirie, Gordon. 2009a. *Air Empire: British Civil Aviation, 1919–39*. Manchester: Manchester University Press.

Pirie, Gordon. 2009b. "British Air Shows in South Africa, 1932/33: 'Airmindedness,' Ambition and Anxiety." *Kronos: Southern African Histories* 35:48–70.

Pirie, Gordon. 2009c. "Incidental Tourism: British Imperial Air Travel in the 1930s." *Journal of Tourism History* 1:49–66.

Pirie, Gordon. 2011. "Non-urban Motoring in Colonial Africa in the 1920s and 1930s." *South African Historical Journal* 63 (1): 59–81.

Prakash, Gyan. 1999. *Another Reason: Science and the Imagination of Modern India*. Princeton: Princeton University Press.

Rabinbach, Anson. 1990. *The Human Motor: Energy, Fatigue, and the Origins of Modernity*. Berkeley: University of California Press.

Rajan, Ravi. 1998. "Imperial Environmentalism or Environmental Imperialism? European Forestry, Colonial Foresters and the Agendas of Forest Management in British India 1800–1900." In *Nature and the Orient: The Environmental History of South and Southeast Asia*, ed. Richard H. Grove, Vinita Damodaran, and Satpal Sangwan, 324–371. Delhi: Oxford University Press.

Ranger, Terence. 1999. *Voices from the Rocks: Nature, Culture and History in the Matopos Hills of Zimbabwe*. Oxford: James Currey.

Rasmussen, R. K., and S. C. Rubert. 1990. *A Historical Dictionary of Zimbabwe*. Metuchen, NJ: Scarecrow Press.

Reuss, Martin. 2008. "Seeing Like an Engineer: Water Projects and the Mediation of the Incommensurable." *Technology and Culture* 49:531–546.

Rodney, Walter. 1972. *How Europe Underdeveloped Africa*. Dar-es-Salaam. Tanzania Publishing House.

Sassen, Saaskia. 1999. *Globalization and Its Discontents*. New York: New Press.

Schama, Simon. 1995. Landscape and Memory. New York: Vintage.

Sheldon, Kathleen E. 2002. *Pounders of Grain: A History of Women, Work, and Politics in Mozambique*. Portsmouth, NH: Heinemann.

Sheller, Mimi. 2011a. "Mobility." *Sociopedia.isa*, 1–12. Available at http://www .sagepub.net.

Sheller, Mimi. 2011b.. "Cosmopolitanism and Mobilities." In *The Ashgate Research Companion to Cosmopolitanism*, ed. M. Nowicka and M. Rovisco, 561–589. Aldershot: Ashgate.

Sheller, Mimi, and John Urry, eds. 2006. *Mobile Technologies of the City*. London and New York: Routledge.

Silla, Eric. 1998. *People Are Not the Same: Leprosy and Identity in Twentieth-Century Mali*. Portsmouth: Heinemann.

Simone, AbdouMalique. 2004. "People as Infrastructure: Intersecting Fragments in Johannesburg." *Public Culture* 16 (3): 407–429.

Smaldone, Joseph P. 1972. "Firearms in the Central Sudan: A Revaluation." *Journal of African History* 13 (4): 591–607.

Smith, Alan. 1969. "The Trade of Delagoa Bay as a Factor in Nguni Politics, 1750–1835." In *African Societies in Southern Africa*, ed. Leonard Thompson, 165–189. London: Heinemann.

Spenceley, Anna. 2008. *Responsible Tourism: Critical Issues for Conservation and Development*. London: Earthscan.

Spinage, C. A. 2003. *Cattle Plague: A History*. New York: Kluwer.

Stiglitz, Joseph. 2003. *Globalization and Its Discontents*. New York: W. W. Norton & Co.

Storey, William K. 2008. *Guns, Race, and Power in Colonial South Africa*. Cambridge: Cambridge University Press.

Sugal, Cheri. 1997. "Elephants of Southern Africa Must Now Pay Their Way." *World Watch* 10:9.

Tapela, Barbara, Lamson Maluleke, and Clapperton Mavhunga. 2007. "New Architecture, Old Agendas: Perspectives on Social Research in Rural Communities Neighbouring the Kruger National Park." *Conservation & Society* 5 (1): 60–87.

Tevera, Daniel S., and Lovemore Zinyama. 2002. *Zimbabweans Who Move: Perspectives on International Migration in Zimbabwe.* SAMP Migration Policy Series No. 25. Cape Town, South Africa: Southern African Migration Project.

Thrift, Nigel. 2004. "Movement-Space: The Changing Domain of Thinking Resulting from the Development of New Kinds of Spatial Awareness." *Economy and Society* 33 (4): 582–604.

Tilley, Helen. 2011. *Africa as a Living Laboratory: Empire, Development, and the Problem of Scientific Knowledge, 1870–1950.* Chicago: Chicago University Press.

Trevor-Roper, Hugh. 1969. "The Past and Present: History and Sociology." *Past & Present* 42:3–17.

Urry, John. 2000. "Mobile Sociology." *British Journal of Sociology* 51 (1): 185–203.

van Onselen, Charles. 1972. "Reactions to Rinderpest in Southern Africa, 1896–97." *Journal of African History* 13 (3): 473–488.

Vansina, Jan. 1986. *Oral Traditions as History.* Madison: University of Wisconsin Press.

Virilio, Paul. 1997. *The Open Sky.* London: Verso.

Voß, Alexander, Rob Procter, and Robin Williams. 2000. "Innovation in Use: Interleaving Day-to-Day Operation and Systems Development." In *PDC 2000: Proceedings of the Participatory Design Conference,* ed. T. Cherkasky, J. Greenbaum, P. Mambrey, and J. K. Pors, 192–201. New York: Computer Professionals for Social Responsibility.

Von Braun, J., T. Teklu, and P. Webb. 1998. *Famine in Africa: Causes, Responses, and Prevention.* Baltimore: Johns Hopkins University Press.

Voss, A. E. 1982. "Thomas Pringle and the Image of the 'Bushmen.'" *English in Africa* 9 (1): 15–28.

Wagner, Roger. 1987. "Zoutpansberg: The Dynamics of a Hunting Frontier, 1848–67." In *Economy and Society in Pre-Industrial South Africa,* ed. Shula Marks and Anthony Atmore, 314–349. London, New York: Longman.

Walton, John. 2006. "Transport, Travel, Tourism and Mobility: A Cultural Turn?" *Journal of Transport History* 27 (2): 129–134.

Thiong'o, Wa. 1986. *Ngugi. Decolonising the Mind.* London: James Currey.

Williams, Denis. 1974. *Icon and Image: A Study of Sacred and Secular Forms in African Classical Art.* London: Allen Lane.

Wolmer, William. 2007. *From Wilderness Vision to Farm Invasions: Conservation and Development in Zimbabwe's South-East Lowveld.* London: James Currey.

Woolgar, Steve. 1991. "Configuring the User: The Case of Usability Trials." In *A Sociology of Monsters: Essays on Power, Technology and Domination,* ed. John Law, 57–102. London: Routledge.

Wright, Marcia. 2002. "Life and Technology in Everyday Life: Reflections on the Career of Mzee Stefano, Master Smelter of Ufipa, Tanzania." *Journal of African Cultural Studies* 15 (1): 17–34.

Zegeye, Abebe, and Robert Muponde, eds. 2012. *Social Lives of Mobile Telephony.* London: Routledge.

Zhao, Zheng. 1994. "Paving a Way to Eldorado: Roads, Railways and Political Economy in Matabeleland, 1888–1914." PhD dissertation, Yale University.

Index

Printed in the United States
by Baker & Taylor Publisher Services